101 Math Activities

by Trudy Aarons, Francine Koelsch

Games, Gameboards and Learning Centers
for Early Childhood Education
and Special Needs Children

**Communication
Skill Builders** ®
3130 N. Dodge Blvd./P.O. Box 42050
Tucson, Arizona 85733
(602) 327-6021

Other publications by Trudy Aarons and Francine Koelsch available through Communication Skill Builders:

- *101 Language Arts Activities, No. 3053-Y, $11.95*
- *101 Reading Activities, No. 2079-Y, $11.95*

The use of the masculine pronoun form is for the sake of convenience and brevity and is not intended to be preferential or discriminatory.

Copyright © 1981 by:

Communication Skill Builders, Inc. ✳®
3130 N. Dodge Blvd./P.O. Box 42050
Tucson, Arizona 85733
(602) 323-7500

ISBN 0-88450-740-8
Catalog No. 2065

INTRODUCTION

101 Math Activities was written to expand our early childhood activities curriculum. *101 Language Arts Activities*, our first book, and *101 Math Activities* give the teacher an opportunity to supplement and reinforce more completely the objectives in an early childhood curriculum with easy-to-make games.

Several years ago as kindergarten teachers, we participated in an ESEA Title I grant. This involved sending home a weekly "Smiley Bag" project with each child. Limited to commercially bought games and activities, we found they did not fill our students' needs since there were many gaps in their skills that needed development and reinforcement. Using our creativity and embellishing upon ideas from colleagues, we began to make many games that filled the math needs of our students. The games were made from available and inexpensive materials. As word of our games and activities circulated, we were invited to conduct several workshops to teacher and parent groups. These workshops were very successful since all participants were able to produce as many games as time permitted. Requests of written game materials for young children started us writing *101 Language Arts Activities* and *101 Math Activities.*

The basic math game forms are designed for children in kindergarten through second grade. The games in each category are sequenced from the easiest skills to the more difficult skills. Before the game is introduced to the class or the individual child, the skill must be matched to the child's readiness level. Each activity states the title of the game, the objective, the materials used, and directions for making and playing the game.

We hope you have as much fun making the games as your children will have in using them.

ABOUT THE AUTHORS

TRUDY AARONS and FRANCINE KOELSCH are teachers in the East Hartford, Connecticut, school system. They are co-authors of Make-A-Game Workshop and have conducted workshops throughout southern New England for parents and educators since 1973.

Ms. Aarons and Ms. Koelsch have authored *101 Language Arts Activities* and have edited and revised *Peel & Put® Reading Program Activity Manual*, both titles available through Communication Skill Builders, Inc.

Both authors received the Master's degree from Central Connecticut State College. Ms. Aarons is the mother of four daughters. She and her husband reside in West Hartford, Connecticut. Ms. Koelsch lives in Marlborough, Connecticut, with her husband and two sons.

DEDICATION

To the students, parents, and teachers
who, through these activities,
helped us demonstrate that learning is fun.

FOREWORD

Disraeli said that "the secret of success is consistency of purpose." Trudy Aarons and Francine Koelsch have demonstrated a consistency of purpose in organizing mathematics activities for the very young and special needs child. Not only are the activities creative, but they pointedly meet the need for reinforcement with these children. Congratulations to them for an outstanding group of activities, the use of which will undoubtedly be perpetuated during the years to come.

Jane K. Piorkowski, Ph.D.
Director of Mathematics
East Hartford School System
East Hartford, Connecticut

CONTENTS

HOW TO BEGIN

Beginnings can be the hardest part of a project. To help you begin, we've listed the materials that have worked best for us, and some avoidable pitfalls.

Materials

* Folding bristol works best for table gameboards and card games. It is flexible, sturdy, and comes in an assortment of colors and sizes.

* Bristol board is heavier than folding bristol. It is excellent for wall and bulletin board learning stations and for large floor games.

* Water-based marking pens are colorful and will make your games attractive. Never use any permanent types of marking pens. They have an oil base that will eventually cause the colors to "bleed."

* Clear adhesive plastic is a must to cover all your games if you wish them to last any length of time.

* Don't worry if you are not artistic. Pictures for your games can be found in seals and workbooks. There are companies that have language arts pictures; try your speech and language clinician for catalogs. *Peel & Put*® pictures available through Communication Skill Builders are excellent.

Avoidable Pitfalls

* When you make a game with cards, cut the cards apart after you have covered them with plastic.

* Paper-punch a hole in the end of the pointer when making a spinner from bristol board. Place a paper fastener through the hole and push it through the spinner. Loosen up the fastener and it will spin freely.

* Make sure the children have all the necessary entry behaviors in order to play the game. Demonstrate how to play the activity before putting out the game.

* Establish a routine with your class for using the activities. For example: (1) the correct number of players, (2) completing the activity before choosing another, (3) checking with the teacher when the activity is completed, (4) putting away the activity.

* Storage of the activity is important. Each activity should have a specific place it belongs and an appropriate container. It should be visible to the children and within their reach. Pieces of the activity should be kept together.

Classification

Shapes

Attributes

COLOR THE TRAIN

Objective

To identify basic shapes (circle, triangle, square, rectangle)

Materials

1. Bristol board, 9" x 12"
2. Colored marking pens, crayon, ruler, compass
3. Clear adhesive plastic

Making the Game

1. At the top of the bristol board, draw a red circle, a blue triangle, a green square and an orange rectangle.
2. Below, draw a train using the four shapes (see illustration).
3. Cover the sheet with clear adhesive plastic.

Playing the Game

1. The child will identify the shapes and colors at the top of the board.

2. He will use crayons to color the shapes on the train to correspond with the color of the shapes above. (All circles are red, all triangles are blue, all squares are green, all rectangles are orange.)

When the child has completed the activity, the crayon marks can be rubbed off with a dry tissue or cloth.

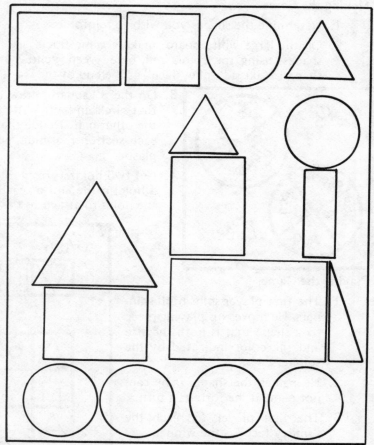

LET'S RACE

Objective

To identify geometric shapes

Materials

1. White bristol board, 18" x 24" and 8" x 8"
2. Folding bristol
3. Wide and narrow marking pens, ruler, compass, two paper fasteners, paper punch
4. One playing piece for each player
5. Clear adhesive plastic

Making the Game

1. Determine the shapes you wish to reinforce.
2. On the large white board, make a game track (see illustration). In random order draw in shapes, using the colors red, blue, green, yellow, and large and small sizes. Color in the shapes. Make a "smiley face" at the end of the track. Cover the board with plastic.

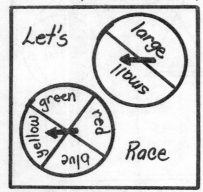

3. On the 8" square, draw two circles 3" in diameter. Divide the first circle in half. Write SMALL on one half and LARGE on the other half. Divide the second circle into four sections. Color each section red, blue, green, or yellow. Cover the square with plastic.
4. Cut two pointers from folding bristol. Cover with plastic. Punch a hole in the end of each, and use the paper fastener through the holes to attach to the spinners.

Playing the Game

1. The first player spins both spinners. He moves his playing piece to a shape that is both the size and the color indicated by the spinners.
2. He names the shape. If he cannot name it, he forfeits a turn.
3. The first player to reach the "smiley face" is the winner.

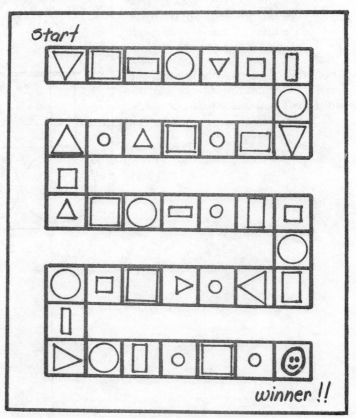

IT'S THE SAME DIFFERENCE

Objective

To identify a shape regardless of its size or arrangement

Materials

1. White bristol board, 18" x 12" and 8" x 8"
2. Folding bristol
3. Pencil, ruler, compass, paper fastener, paper punch
4. One playing piece for each player
5. Clear adhesive plastic

Making the Game

1. On the large white board, make a game track (see illustration below). In random order draw in shapes, using different colors, sizes, and arrangements. Cover the board with plastic.
2. Draw a 7" circle on the square bristol. Divide it into five sections. Draw one of each shape and a "sad face" in the sections. Color the shapes black.
3. Make a pointer from folding bristol. Cover it with plastic, punch a hole in one end, and attach it to the center of the circle with the paper fastener.

Playing the Game

1. Name the shapes on the spinner with the players.
2. Each player in turn spins the spinner and moves forward on the board to the first space that matches the shape on the spinner.
3. If the pointer shows the "sad face," the player forfeits a turn.
4. The first player to reach the last space is the winner.

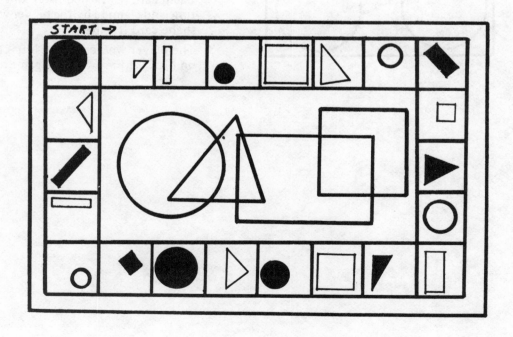

SHAPES IN SPACE

Objective

To recognize common shapes in space

Materials

1. Folding bristol
2. Pictures (four of each) of common items that have the following shapes: cube, sphere, circular cone, circular cylinder
3. Marking pens, compass, paper cutter, scissors, paste
4. Clear adhesive plastic

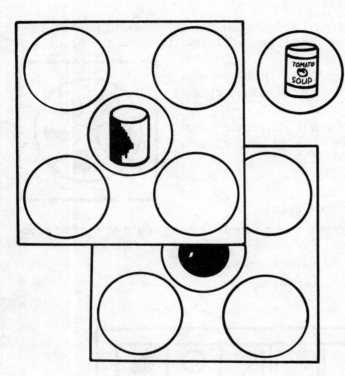

Making the Game

1. Cut four 6" x 6" boards from the folding bristol. Draw five 2" circles on each board. Draw one shape in the center circle of each board. Cover the boards with plastic.

2. Make cue cards by drawing sixteen 2" circles on another sheet of folding bristol. In each circle, paste a picture of an item having each shape. Cover the board with plastic and cut out the circles.

Playing the Game

This is a basic lotto game for four players.

1. The cue cards are placed face down in the center of the playing area.

2. Each card is turned over one at a time and claimed by the player whose shape card it matches.

3. The player who covers all four circles on his card first wins the game.

STRETCH IT

Objective

To duplicate basic shapes

Materials

1. 6" x 6" wooden board
2. Forty-nine push-pin thumb tacks
3. Colored elastics
4. Folding bristol board
5. Marking pens, paper cutter, ruler
6. Clear adhesive plastic

Making the Game

1. Place the push pins in the board to make seven rows of seven pins evenly spaced.
2. Rule the folding bristol into 4" x 4" sections. In each section, draw a basic shape (circle, triangle, square, diamond, etc.). Cover with plastic and cut into cards.

Playing the Game

1. The cards are placed face down on the playing area.
2. The child turns over the top card. Stretching the elastics on the push pin board, he duplicates the shape indicated on the card.
3. Play continues until he has turned over all cards and duplicated all shapes.

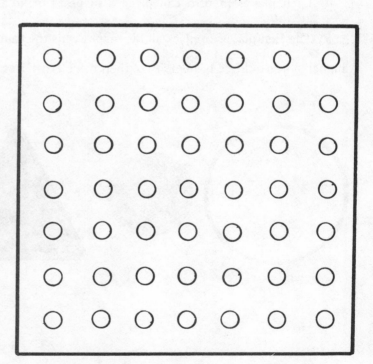

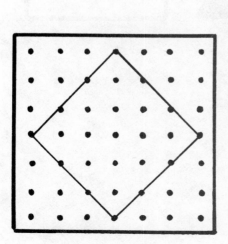

ATTRIBUTE TRAIN

Objective

To match objects by determining common properties

Materials

1. Bristol board (one sheet each of red, yellow, blue, green)
2. Compass, ruler, protractor, pencil, scissors
3. Clear adhesive plastic

Making the Game

1. Make two complete sets of shapes in each color. One set will be large (4" high) and the other will be small (2" high). Each set of shapes will consist of a square, triangle, circle, rectangle, diamond.
2. Cover all shapes with plastic and cut them out.

Playing the Game

1. Spread out the forty shapes. Ask the group to sort them any way they can. By doing so, they will discover the attributes of the set (color, shape, size).
2. Mix the set and spread it out again. Divide the set evenly among the players. Place one shape in the center of the playing area.
3. The first player will place his shape next to the first shape if he can name a common attribute (they are both big, they are both yellow, etc.). If he cannot match and name an attribute, he loses his turn.
4. Each player in turn can place a shape to make a "train" if he can match and name an attribute.
5. The first player to place all his shapes wins the game.

A higher skill would be to name two, then three attributes.

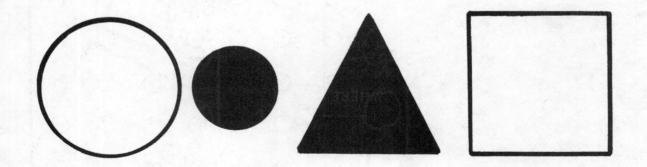

"GUESS" ATTRIBUTE CONCENTRATION

Objective

To match objects by determining common properties

Materials

1. White folding bristol
2. Marking pens, ruler, paper cutter, pencil
3. Clear adhesive plastic

Making the Game

1. Decide on the attributes you wish to reinforce (for example, flowers, shapes, colors, and large and small sizes).
2. Rule the folding bristol into 3" x 3" sections. Make two sets of flowers and shapes in every color and in large and small sizes. Cover the cards with plastic and cut apart.

Playing the Game

The object is to find two cards that have a common attribute (they are both large, they are both flowers, etc.).

1. Place the cards face down in rows on the playing area.
2. The first player turns over two cards. If they match, he takes them and turns over two more cards, continuing until he cannot find a matching pair. If the cards do not match, he turns both cards face down, and the next player takes his turn.
3. The player with the most pairs is the winner.

To increase the difficulty of the activity, have the player name two common attributes (they are both small and blue, etc.).

"MATCH IT" DOMINOES

Objective

To match objects by determining common properties

Materials

1. White bristol board
2. Marking pens (red, blue, green, yellow, black)
3. Ruler, pencil, paper cutter
4. Clear adhesive plastic

Making the Game

1. Determine the attributes you wish to reinforce (for example: large, small, colors, flowers, houses, shapes).
2. Make dominoes by ruling the bristol board into 2½" x 5" sections. Divide each section in half with a bold black line. Draw flowers, houses and shapes, using the four colors and making them large and small. Cover with plastic and cut apart into individual dominoes.

Playing the Game

1. Deal five cards to each player. Place the remaining cards face down in the middle of the playing area.
2. Turn over the top card. Discuss the attributes of the picture.
3. The first player matches a card to the center card. He names the common attribute (they are both red).
4. Each player in turn plays a card, naming the common attribute.
5. If the player cannot play a matching card, he draws from the deck until he finds a matching card.
6. The first player to play all his cards wins the game.

A more difficult skill is to name two or three attributes.

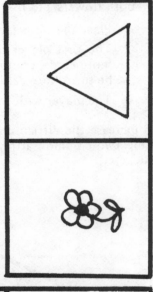

MAKE A GUESS

Objective

To describe an unseen object

Materials

1. White folding bristol
2. Bristol board, 22" x 28"
3. Pencil, ruler, paper cutter, paper punch
4. Marking pens (yellow, green, blue, red, black)
5. Sixteen small drapery hooks
6. Clear adhesive plastic

Making the Game

1. Decide on the common attributes you wish to reinforce (for example, four shapes and four colors).
2. Rule the folding bristol into sixteen 4" x 6" sections. In each section, draw a shape in each color. Cover with plastic, punch a hole at the top of each section, and cut apart into cards.
3. Rule the bristol board into sixteen sections, following the illustration. Use bold black lines. Cover with plastic and insert a hook in the top of each section.
4. Hook the cards to the board vertically according to color, and horizontally according to shape. All colors and shapes will be shown.

Playing the Game

1. Randomly turn over four or five cards, concealing the shapes and colors.
2. The player must guess the card that is hidden and state why he has made his decision.

To increase the difficulty of the game, decrease the number of cards seen.

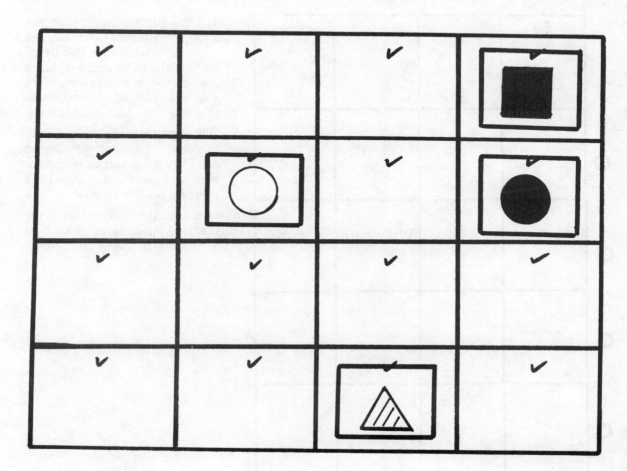

DRAW IT FAST!

Objective

To place an object by determining its properties

Materials

1. Duplicating master
2. Ruler, pencil
3. Crayons

Making the Game

1. Mark off a 4" x 5" rectangle on the duplicating master.
2. Rule the rectangle into twenty sections (four across, five down) to make a gamesheet master.
3. Make copies of the gamesheet.

Playing the Game

1. Decide on the attributes you wish to reinforce. In this example, we have used color and directionality. (You may adapt the game to reinforce size, shape, classification, texture, etc. — the combinations of attributes are limited only by the fine motor skills of the group.)
2. Give each student a gamesheet.
3. Tell the students to use their crayons to place a different color dot in the left margin next to each of the bottom four horizontal sections. Tell them to work from the bottom up. The bottom color will be blue, the next yellow, the third color will be green, and the next red. Tell the group that any color placed in the row must be the same color as the dot in the margin.

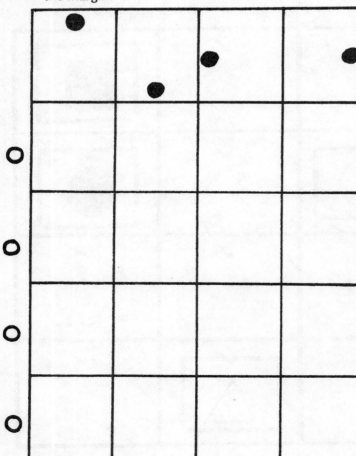

4. Tell the group that a square has a top, a bottom, a right side, and a left side. Reinforce this by directing them to place a black dot at the top, bottom, right and left sides of the squares in the top row (see illustration).
5. Tell them to place a green left-side dot, a blue top dot, and so on. They will practice directionality within vertical columns and color by horizontal rows.
6. Continue the activity until the students have placed color dots in all squares.

Ordering

REPEAT, REPEAT

Objective

To recognize and repeat a pattern

Materials

1. Bristol board
2. Marking pens, paper punch, paper cutter, paper fasteners
3. Clear adhesive plastic

Making the Game

1. Rule the bristol into 2½" x 12" sections. On each section, rule 2" squares.

2. Use one of the cards to make a pattern card, drawing shapes in each square. Use a simple two-color pattern (red circle, blue square, repeated). Leave the two end squares blank (see illustration).

3. Using another card, make a more difficult pattern card (red circle, blue square, green triangle), leaving three end squares blank.

4. Make additional pattern cards, varying the patterns and difficulty.

5. Using the remaining cards, in each square draw a shape to match the shapes and colors on the pattern cards. Be sure to make as many shape cards as necessary to complete all blanks on the pattern cards.

6. Cover all cards with plastic.

7. On the pattern cards, punch a hole at the side of each blank space (see illustration).

8. On the remaining cards, punch a hole at the top of each and cut into individual 2" cards. Attach a paper fastener through each hole.

Playing the Game

1. The player will take a pattern card and complete the pattern by attaching the individual squares in correct sequence.

2. Play continues until all 2" cards are used to complete the patterns on all the pattern cards.

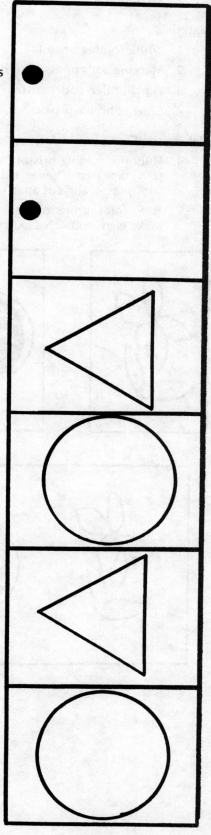

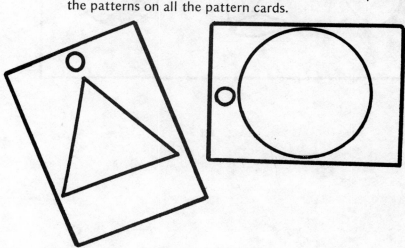

COPY ME

Objective

To duplicate a pattern

Materials

1. White folding bristol
2. Marking pens or assorted duplicate sets of small pictures and paste
3. Pencil, ruler, paper cutter
4. Clear adhesive plastic

Making the Game

1. Rule the folding bristol into 3" x 9" cards. On each card, draw a simple two-color pattern, repeated (flower, ball, flower, ball); or paste small pictures in sets. Cover the bristol with plastic and cut apart.
2. Rule folding bristol into 2" x 3" cards. For every design or picture on a pattern card, make eight to twelve matching cards. Cover with plastic and cut apart.

Playing the Game

1. Give each player a pattern card and a set of small pictures that match.
2. The players will duplicate the pattern, using the small cards.

Higher skills include extending the pattern and duplicating the pattern from memory.

BULLETIN BOARD FUN

Objective

To create a pattern and extend it

Materials

1. Bristol board, 22" x 28"
2. Folding bristol
3. Ruler, paper cutter, paste
4. Small drapery hooks
5. Sets of duplicate pictures (six of each)
6. Clear adhesive plastic

Making the Game

1. Rule the bristol board into three strips, 7" x 28". Cover with plastic and cut apart. Place the strips end to end across the bulletin board. Insert drapery hooks in the strips, spaced about 6" apart.
2. Rule the folding bristol into 4" x 6" cards. Paste sets of pictures on the cards. Cover with plastic and cut apart. Punch a hole in the top of each card.

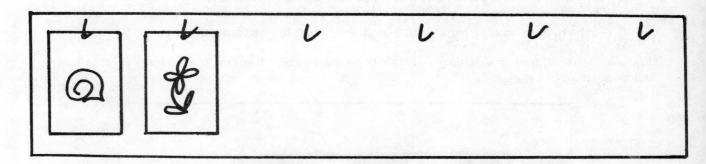

Playing the Game

1. Show the group how to hang the pictures on the hooks, from left to right. Start with a two-picture pattern.
2. The child will duplicate the pattern across the bulletin board, attaching the cards through the hooks.

Increase the difficulty of the pattern according to your group's needs, or have the children create their own patterns. This is an excellent independent activity.

COLOR THEM ALL

Objective

To extend a given pattern

Materials

1. White folding bristol
2. Duplicating masters
3. Marking pens, crayons, pencil, ruler
4. Clear adhesive plastic

Making the Game

1. Rule the folding bristol into 3" x 9" pattern cards. Draw a row of six 1" squares in the center of each card (see illustration). Color the squares, starting with simple two–color patterns and increasing the difficulty to fit your group. Cover the cards with plastic and cut apart.

2. Rule the duplicating masters into 4" x 11" cards. In the center of each, make a row of nine 1" squares.

Playing the Game

1. Give each player a duplicating master. Choose a pattern card. Show the card to the group. Discuss the pattern.

2. Each child will color his paper to match, extending the pattern on his paper.

For a higher skill, flash the pattern and have the children color from memory. This is an excellent independent activity.

WHERE DO WE GO?

Objective

To sort a collection into sets

Materials

1. Bristol board, 22" x 28"
2. Folding bristol
3. Thirty drapery hooks
4. Assortment of small pictures, six each in six categories
5. Marking pen, ruler, pencil, paper cutter, paper punch
6. Clear adhesive plastic

Making the Game

1. Following the illustration, rule the bristol board into six columns. Place one picture from a set at the top of each column. Draw a different colored line under each picture. Cover the board with plastic.
2. Place five drapery hooks in each column, evenly spaced.
3. Rule the folding bristol into 3" x 3" cards. Place the remaining pictures on the cards. Color code each set of cards to match the color line on the board. Cover with plastic and punch a hole in the top of each card.

Playing the Game

1. Name the pictures on the board with the players.
2. Spread out the cards on the playing area. Choose one card and ask the child where it belongs on the board.
3. The child will find all the cards that belong to that set. Then he will do the same for the remaining sets. The color on the back of the cards should match the line.

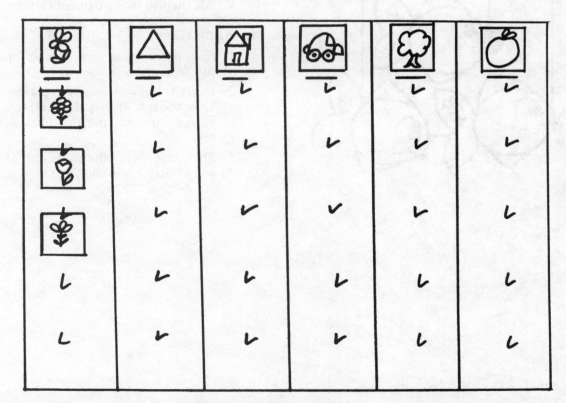

GONE FISHING

Objective

To match sets of objects

Materials

1. A cardboard carton
2. Paint
3. One dowel
4. String
5. A magnet
6. Construction paper in a variety of colors
7. Stickers
8. Scissors, paper clips
9. Clear adhesive plastic

Making the Game

1. Paint the cardboard carton.
2. Tie a string on the stick, then tie a magnet to the end of the string.
3. Draw twelve sets of fish from construction paper.
4. Put sets of stickers on the fish, making matches of sets on several fish.
5. Cover with clear adhesive plastic and cut out the individual fish.
6. Place one paper clip on each fish.
7. Put the fish in the painted carton.

Playing the Game

1. Each player takes turns fishing for two fish from the box.
2. If the fish match sets, the player may keep the fish. If they don't match, the player must put the fish back in the box.
3. The player with the most fish at the end of play is the winner.

WHERE'S MINE?

Objective

To match the elements of one set with those of a second set

Materials

1. Bristol board, 22" x 28"
2. Folding bristol in contrasting color
3. Twenty-four drapery hooks
4. Thirty-two small pictures of items that "go together," eight of each item (for example, Set 1: eight bats and eight balls; Set 2: eight apples and eight oranges)
5. Marking pen, ruler, pencil, paper cutter, paper punch
6. Clear adhesive plastic

Making the Game

1. Divide the bristol board into two sections. Place eight identical members of one set on one side of the board. Draw a dotted line to the other side of the same section. Do the same in the second section, using eight identical pictures from the second set. Cover the board with plastic. Insert drapery hooks at the end of each dotted line.
2. Rule the folding bristol into 2½" x 2½" cards. Place the remaining pictures on the cards. Cover with plastic and cut apart. Punch a hole at the top of each card.

Playing the Game

This is an activity for two players.

1. Each player chooses one side of the board. The small cards are mixed and placed face down on the playing area.
2. The first player turns over a card. If the picture is one that "goes together" with the set on his side of the board, the player hangs it on the first hook. If it does not "go together" with his set, he replaces it face down and the second player has a turn.
3. The first player to find all of his cards is the winner.

Reinforce the term "matching one-to-one."

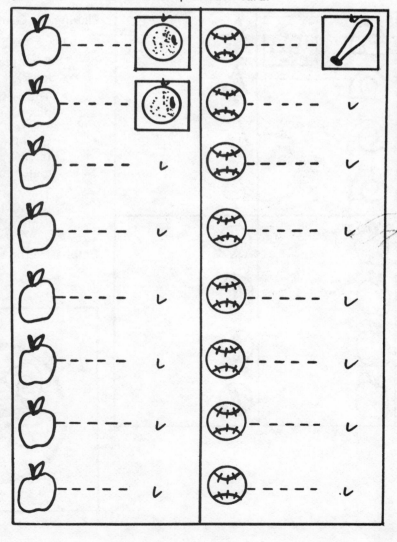

MATCH THE SET

Objective

To make equivalent sets

Materials

1. Bristol board, 12" x 18" and 7" x 7"
2. Folding bristol
3. Assortment of thirty–six small pictures (nine identical pictures in four categories)
4. Marking pen, ruler, paper cutter, paper punch, paper fastener, compass
5. Clear adhesive plastic

Making the Game

1. Rule the bristol board into four sections. Place four identical pictures in each section. Cover the board with plastic.
2. Rule a piece of folding bristol into 2½" x 2½" cards. Use four pictures from each set and place them on the cards. Cover the cards with plastic and cut apart.
3. Using the compass, draw a 6" circle on the bristol square. Rule the circle into four sections. Place a picture from each set into the sections. Cover with plastic.
4. Make a pointer from folding bristol. Cover with plastic, punch a hole in the end, and attach it to the center of the circle with a paper fastener.

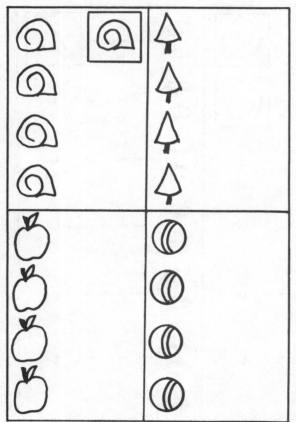

Playing the Game

This is a game for two to four players.

1. Each player chooses a set of four matching 2½" cards.
2. The first player spins the spinner. If the picture on the spinner matches the player's set, he places one card on the corresponding set section on the gameboard. If the picture on the spinner does not match his set, he forfeits a turn.
3. The first player to match all of his pictures with those on the board wins the game.

Emphasize the term "matching one-to-one."

PUNCH OUT THE MATCH

Objective

To match according to numbers and shapes

Materials

1. Four colors of bristol board
2. Paper punch, scissors
3. Clear adhesive plastic

Making the Game

1. On the bristol board, make shapes (circle, triangle, square, rectangle). Make four of each color and shape. Cover with plastic and cut out.
2. On each shape, punch from one to ten holes, making matches in the various shapes (see illustration).

Playing the Game

1. The child will match the shapes according to the number of holes punched (all twos together, etc.).
2. He then makes matches according to shapes (all circles go together, etc.).

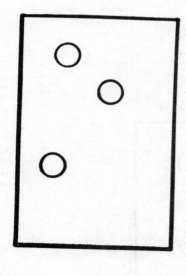
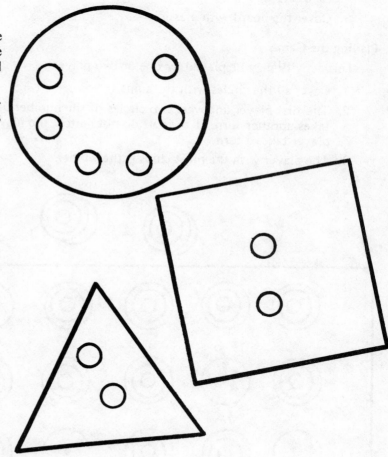

SET CHIP CONCENTRATION

Objective

To match equivalent sets despite the arrangement of the elements in each set

Materials

1. Folding bristol, 12" x 18"
2. Twenty poker chips
3. Marking pens
4. Clear adhesive plastic

Making the Game

1. Draw twenty circles on the bristol board by tracing around a chip.
2. In the circles, draw four sets of from one to five dots. Vary the arrangement of dots in each set.
3. Cover the board with plastic.

Playing the Game

This activity may be played by one or two players.

1. Cover all the circles with the chips.
2. The first player uncovers two circles. If the number of dots match, he keeps the chips and takes another turn. If the sets do not match, the player replaces the chips and the second player takes a turn.
3. The player with the most chips is the winner.

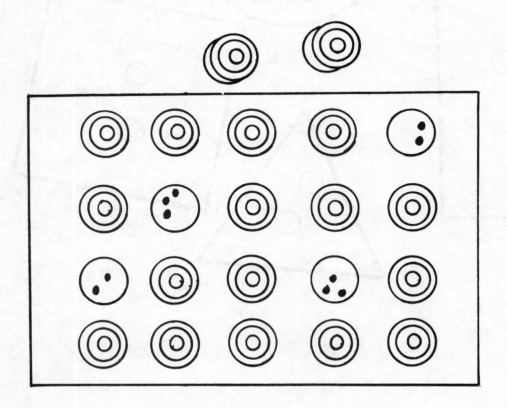

SET DOMINOES

Objective

To match equivalent sets despite the arrangement of the elements in each set

Materials

1. Bristol board
2. Marking pens
3. Paper cutter
4. Clear adhesive plastic

Making the Game

1. Make dominoes by ruling the bristol board into 2½" x 4" sections. Divide each section in half. Place sets of dots of from one to five on each section. Vary the arrangement of dots for each set.
2. Cover with plastic and cut apart into individual cards.

Playing the Game

1. A double card (one with the same set on each end) is placed in the center of the playing area. The cards are shuffled and dealt, seven to each player. The rest of the cards are placed face down in a pile.
2. Each player in turn matches one of his cards to the ends of the cards. If he cannot play a card, the player draws from the deck until he gets a card that matches one of the ends.
3. The first player to play all of his cards is the winner.

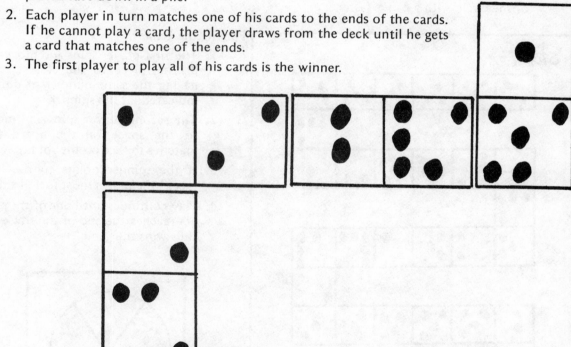

CONSERVATION CHASE

Objective

To recognize the number of a set despite the arrangement of its elements

Materials

1. Bristol board, 18'' x 24'' and 8'' x 8''
2. Marking pens, ruler, paper fastener, paper punch
3. One playing piece for each player
4. Clear adhesive plastic

Making the Game

1. Make a game track on the bristol (see illustration). Thirty to thirty-six spaces makes an interesting game.
2. Make sets of from one to five dots, repeating them in random order around the track. Use the colored marking pens randomly. Cover the board with plastic.
3. Draw a 7'' circle on the bristol board square. Divide the circle into six sections. Make sets of from one to five dots in the five sections, and draw a "sad face" in the sixth section. Cover the board with plastic.
4. Make a pointer, cover it with plastic, punch a hole in one end, and attach it to the center of the circle with a paper fastener.

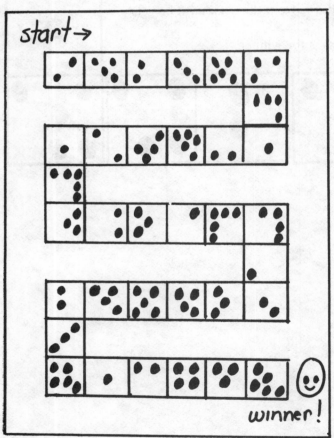

Playing the Game

1. The first player spins and moves his playing piece to the closest space having the same number of dots as indicated on the spinner.
2. The second player spins and moves to the space on the board that matches the set on the spinner.
3. If the spinner points to the "sad face," the player must forfeit a turn.
4. Play continues until one of the players reaches the end of the track and thus wins the game.

CHIPS AWAY

Objective

> To recognize the number of a set despite the arrangement of its elements

Materials

1. Bristol board, 18" x 24" and 8" x 8"
2. Thirty-six poker chips
3. Marking pens, ruler, paper fastener, paper punch, compass
4. Clear adhesive plastic

Making the Game

1. Make thirty-six circles on the bristol board with the compass. Make sets of from one to five dots in random order in the circles. Color the sets randomly. Cover the board with plastic.
2. Draw a 7" circle on the bristol square. Divide the circle into six sections. Make sets of from one to five dots in five of the sections and a "sad face" in the sixth section. Cover with plastic.
3. Make a pointer, cover it with plastic, punch a hole in one end, and attach it to the center of the circle with the paper fastener.

Playing the Game

1. Divide the poker chips equally among the players.
2. The first player spins the spinner. He places one of his chips on any space having a number of dots matching those on the spinner.
3. Each player in turn spins the spinner and places a chip on any matching space. If the spinner points to the "sad face," the player must forfeit his turn. A turn also is forfeited if all the matching spaces are covered.
4. The first player to place all his chips on the board wins the game.

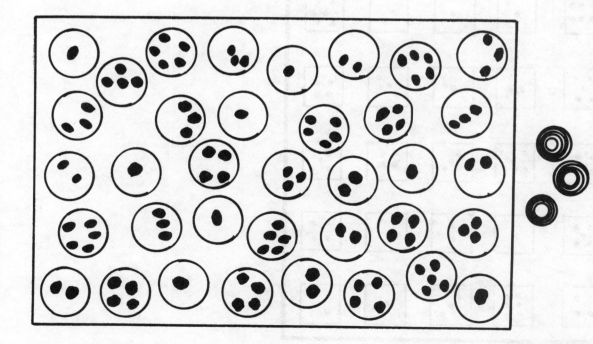

DOT BINGO

Objective

To recognize identical dots on a die and in a square

Materials

1. Folding bristol, 9" x 12"
2. Marking pen, ruler
3. Two sets of colored chips, a die
4. Clear adhesive plastic

Making the Game

1. Rule off a 2" space at the top of the folding bristol. Print "Dot Bingo" in the space.
2. On the rest of the folding bristol make 1" x 1" boxes in rows, five boxes across and five boxes down.
3. In each box make dots from one to six.

Playing the Game

This is a game for two players.

1. Each player chooses chips of one color.
2. The first player throws the die. The player then puts one chip on the square that shows the same number of dots.
3. The second player proceeds to throw the die and places a chip on the corresponding square.
4. Players try to block each other from covering one line either horizontally, vertically or diagonally.

5. Play proceeds until one player covers an entire row with his chips.

BIG TO SMALL PUZZLE

Objective

To recognize sizes from the largest to smallest

Materials

1. Bristol board
2. Colored marking pens, scissors
3. Clear adhesive plastic

Making the Game

1. Mark off six 6" x 8" cards.
2. On each card draw a tree, starting with a large tree and graduating to a small tree.
3. Cut puzzle shapes on the edges of the cards so they will fit together from the largest tree to the smallest. Cover the cards with plastic. Cut along the puzzle shapes.

Playing the Game

1. The child will look at the cards and decide which is the largest drawing. He will start the puzzle with that piece.
2. He then will select the drawing that is the next size. If this card fits the puzzle shape, it is correct.
3. The child will complete the puzzle, arranging the drawings from the largest to the smallest.

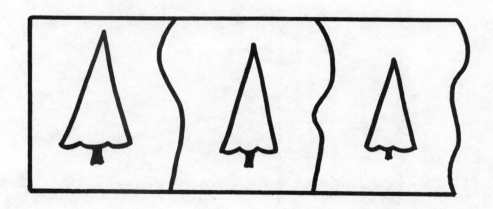

ALL ABOARD

Objective

To determine the sequence of a given set

Materials

1. Bristol board, two 9" x 12" pieces
2. Folding bristol
3. Marking pens, pencil, ruler, paper cutter, scissors
4. Clear adhesive plastic

Making the Game

1. Decide what you wish to seriate (sizes, shapes, colors, numbers).
2. Following the illustration, draw an engine on one bristol board. Draw the caboose on the other bristol board. Decorate the train. Cover with plastic and cut out the train pieces.
3. Rule the folding bristol into 4" x 6" sections. In each section, draw the objects to be seriated. Cover the folding bristol with plastic and cut apart into individual cards.

Playing the Game

1. Place the engine and the caboose on the playing area. Spread out the cards.
2. Tell the child that he is to make the train bigger by putting the cards in order.

For a higher skill, the child will determine the sequence without clues from the teacher.

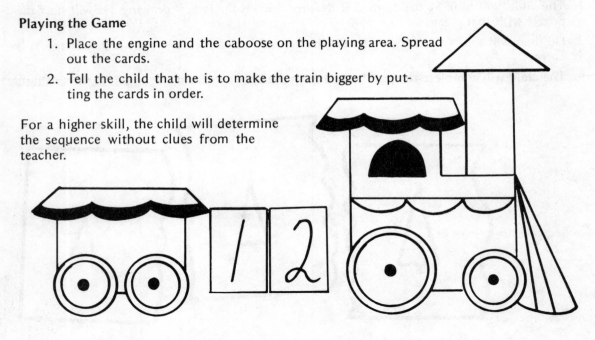

CLOTHESLINE RACE

Objective

To determine the sequence of a given set

Materials

1. Bristol board, 22" x 28"
2. Folding bristol
3. Marking pens, pencil, ruler, paper cutter, paper punch
4. Small drapery hooks
5. Clear adhesive plastic

Making the Game

1. Decide what you wish to seriate (sizes, shapes, colors, numbers).
2. Cut the bristol board into two pieces, each measuring 11" x 28". Draw a clothesline in each section (see illustration). Decorate. Cover with plastic.
3. Rule the folding bristol into 3" x 4" sections. On each section draw the objects for seriation. Cover with plastic and cut apart into individual cards. Punch two holes in the top corners.
4. Place the drapery hooks along the clothesline to correspond with the holes in the cards.

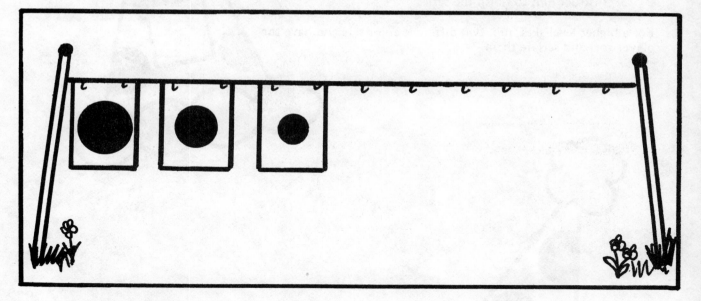

Playing the Game

This is a game for two players.

1. Give each player a "clothesline" and a set of cards.
2. The first player to hang all his cards in order on the line wins the game.

This is an excellent activity to reinforce seriation on an individual basis.

COLOR CIRCUS

Objective

To put colors in order from lightest to darkest, and from darkest to lightest

Materials

1. White folding bristol
2. Paint chips (samples from hardware store)
3. Marking pens, pencil, ruler, scissors
4. Clear adhesive plastic

Making the Game

1. Following the illustration, make clown hats that measure 3"x 4". Use the marking pens to decorate the hats.
2. Glue a paint chip in the center of each hat. Cover with plastic and cut out with scissors.

Playing the Game

1. Spread out the hats in the center of the playing area.
2. Help the player by selecting the lightest color.
3. Instruct him to finish the series.

For a higher level skill, mix two different color sets and have the player sort and seriate them.

DOTS ARE FUN

Objective

To determine the sequential order of a given set of numerals

The advantage of making your own dot-to-dot cards is that you may start the numerical sequence with any number you wish. Make several cards beginning with different numbers.

Materials

1. Folding bristol, 9" x 12"
2. Coloring book of simple objects
3. Marking pens, crayons, pencil, eraser
4. Clear adhesive plastic

Making the Game

1. Using the pencil, lightly trace a picture of an object onto the folding bristol.
2. With the marking pen, make dots and write numerals around the outline of the object. Add details with the markers.
3. Erase the pencil marks. Add details with the markers.

Playing the Game

Have the child connect the dots, using a crayon.

When the activity is completed, the marks can be wiped off with a dry tissue or cloth.

For a higher skill, have the child connect the dots, working backward.

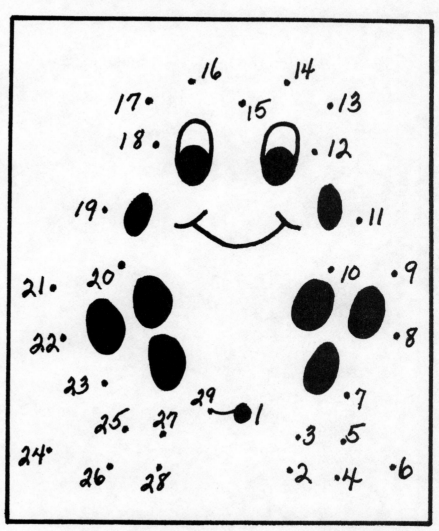

Counting

WINNIE THE POOH WALK

Objective

To reinforce one-to-one association of numbers one to six

Materials

1. White bristol board, 18" x 24"
2. Four pictures of Winnie the Pooh
3. Marking pens, paste
4. One die, a playing piece for each player
5. Clear adhesive plastic

Making the Game

1. Paste a picture of Winnie the Pooh in each corner of the bristol board.
2. Starting at each picture, draw a path around the gameboard, using 1" squares (see illustration).
3. Cover the gameboard with clear plastic.

Playing the Game

This is a game for four players.

1. Each player chooses one of the pictures of Winnie the Pooh as his starting point.
2. The first player tosses the die. Starting at his picture, he moves the same number of spaces as the die shows.
3. Play continues in turn until one player has won the game by moving around the board and returning to his starting point.

DOTTIE DOG

Objective

To recognize numbers from one to ten

Materials

1. Bristol board, 24" x 36"
2. Black marking pen, scissors
3. Clear adhesive plastic

Making the Game

1. Draw a large Dalmatian dog (see illustration).
2. Divide the dog into eleven puzzle pieces. Label each piece from zero to ten, from the nose (0) to the hind leg and tail (10). Draw dots on each piece to correspond with the number.
3. Cover the dog with plastic. Cut out each puzzle piece.

Playing the Game

1. The child looks at each piece and names the number.
2. He then puts the dog puzzle together in numerical sequence.

FOOTPRINT WALK

Objective

To identify and name numbers one through ten

Materials

1. Bristol board
2. Marking pen, scissors
3. Clear adhesive plastic

Making the Game

1. Trace ten footprints on the bristol board.
2. Write a numeral (1 to 10) on each footprint. Draw a corresponding number of dots on each footprint.
3. Cover with plastic and cut out the individual footprints.

Playing the Game

1. The child spreads the footprints on the floor in random order.
2. He steps on each footprint and names the number stepped on. If he misses naming the number, he must start again from the beginning.
3. The child then places the footprints on the floor in numerical order.

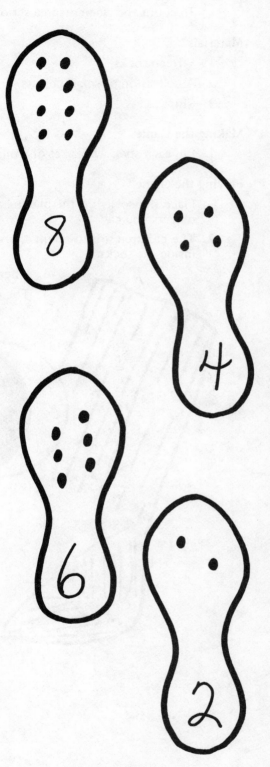

42

FIND THE SOCK

Objective

To count the elements in a set of from one to five unseen objects

Materials

1. Fifteen socks
2. A collection of small objects
3. Chips

Making the Game

1. Fill each sock with a set of from one to five objects.

Playing the Game

1. Place the socks on the playing area. Ask the children to find the sock with the number of objects you call.
2. The children are to count the sets only by feeling the objects when they put their hands inside the socks.
3. When the child finds the correct sets, he is given a chip.
4. The player with the most chips is the winner.

LUNCH TIME

Objective

    To compare the elements of two sets to determine if they are equivalent

Materials

    1. Ten lunch bags

    2. Collection of small objects

Making the Game

    1. Place one to five objects in each of five bags. Duplicate the sets so you have two of each kind of bag.

Playing the Game

    This is a game for ten players. The object of the activity is to find the person who has a bag with the same number of objects as your bag.

    1. Each player selects a bag and chooses a partner.

    2. The partners compare objects to see if the sets are equal. If the sets are unequal, the players keep changing partners until they find a matching pair.

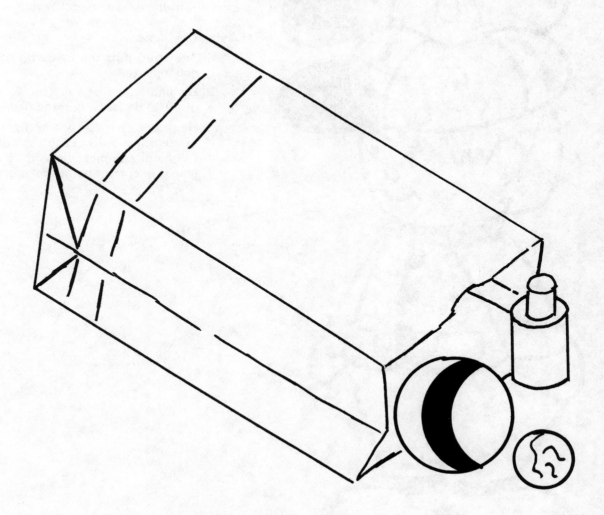

CAPS FOR SALE

Objective

To count by twos, fives, or tens

Materials

1. Colored bristol board
2. Marking pens, scissors
3. Clear adhesive plastic

Making the Game

1. Draw a figure on bristol board (see illustration). Cover with plastic and cut out.
2. Draw ten caps of various colors. Mark each cap with a number (multiples of two, five, or ten).
3. Cover with plastic. Cut out the individual caps, cutting each top into a puzzle shape to fit the cap with the next consecutive number.

Playing the Game

1. The child puts the figure on the playing area.
2. He finds the first cap (the 2, 5, or 10) and places it on the figure.
3. He places each cap on the figure in counting order. Each cap will fit the puzzle piece beneath it if placed in correct numerical order.

GINGERBREAD COOKIES

Objective

    To solve a simple puzzle involving ordinal positions from first to fifth

Materials

1. Brown folding bristol
2. Marking pens
3. Clear adhesive plastic

Making the Game

1. Cut out a free form from the bristol board.
2. On the form, draw five small gingerbread shapes.
3. On each gingerbread shape, make one to five dots representing buttons.
4. Cover with clear adhesive plastic.
5. Cut out the gingerbread shapes from the large piece, being careful not to tear the puzzle shape.

Playing the Game

1. The child looks at each gingerbread shape and identifies the number of dots.
2. He places the one-dot gingerbread in the first shape, the two-dot gingerbread in the second shape, etc.
3. Play continues until all shapes are placed in correct sequence.

BOXES IN ORDER

Objective

 To recognize the ordinal positions designated by first to fifth

Materials

1. Equal-sized boxes or containers
2. Marking pens
3. Small items (matchbox car, small ball, ribbon, lollipop, etc.)

Making the Game

1. Place the five boxes in a row.
2. Label the boxes from one to five.

Playing the Game

1. Place the items on the table in front of the boxes.
2. Ask the child to identify the numbers on the boxes and name the objects.
3. Give directions to the child (e.g., put the car in the third box, the lollipop in the fifth box, etc.). The child will place the items correctly.

SHOPPER'S DELIGHT

Objective

To identify ordinal numbers first through sixth

Materials

1. Bristol board, 12" x 18"
2. Paper fastener, paper punch, ruler, paste, scissors
3. Sets of colored chips
4. Small pictures
5. Empty margarine container
6. Clear adhesive plastic

Making the Game

1. Draw the outline of a building (a "department store") on the bristol board. On the left side, mark off a 2½" x 3" area (for an "elevator shaft"). Mark off rectangles ("floors") measuring 2½" x 9". Label each level starting from the bottom (1st to 6th).

2. On each level paste five small pictures of the same category (e.g., 1st floor, fruits; 2nd floor, pets; 3rd floor, toys; 4th floor, clothing; 5th floor, sporting goods; 6th floor, furniture).

3. Cut out a circle to fit the cover of a margarine container. Divide the circle into six sections. In each section, paste a picture belonging to one of the six categories. Cover the board and the circle with plastic.

4. Paste the circle on top of the margarine cover. Make a pointer, cover it with plastic, punch a hole at the end, and fasten it to the middle of the margarine cover with a paper fastener. Store the chips in the container.

Playing the Game

This is a game for two or more players.

1. Each player receives ten chips of the same color.
2. The first player spins and identifies the category indicated by the pointer. He then names the floor on which the item can be found. If named correctly, he places a chip on an item on that floor.
3. Play continues with each player taking a turn. If all items on a floor are covered, the player loses a turn.
4. The first player to use all his chips is the winner.

CLOTHESPIN COUNT

Objective

To determine which of two sets has more or fewer elements

Materials

1. Bristol board
2. Folding bristol
3. Ten ziplock plastic bags
4. Fifty-five spring-type clothespins
5. Marking pens
6. Clear adhesive plastic

Making the Game

1. Rule the bristol board into ten 6" x 9" sections. In each section mark from one to ten numerals and a corresponding number of dots. Cover with plastic and cut apart into individual numeral cards.

2. Rule the folding bristol into ten 3" x 4" cards. Number the cards from one to ten. Cover with plastic and cut apart. Place one card and a matching number of clothespins in each bag.

Playing the Game

1. Give the child a large numeral card and a bag of clothespins and numbers.

2. Ask him if the number on the numeral card is more or less than the number written on the card in his bag.

3. After giving the correct answer, the child opens his bag and clips a matching number of clothespins on the dots on the number card.

4. Give the child another numeral card and bag of clothespins. Play continues until he has clipped all pins to all numeral cards.

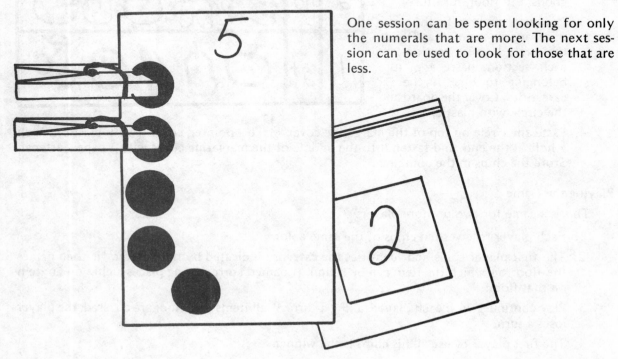

One session can be spent looking for only the numerals that are more. The next session can be used to look for those that are less.

FIND THE SMILEY

Objective

To identify numbers that are greater or less than a specific number

Materials

1. Folding bristol
2. Marking pen, paper cutter
3. Clear adhesive plastic

Making the Game

1. Rule the folding bristol into 4½" x 12" sections. In the center of each section, mark a number and one of the terms *less than* or *greater than*. Cut apart into individual cards.
2. Fold 2" at each end of the cards. On the flaps, mark two numbers: one that corresponds to the middle number and term, and one that does not. Under the flap of the correct number, draw a "smiley face."
3. Cover all rectangle cards with clear plastic. Crease the folds again.

Playing the Game

1. The child will look at the card and read the information in the middle.
2. He will select an answer by turning over one flap. If his answer is correct, there will be a "smiley face."
3. The child continues to find correct answers to all the cards.

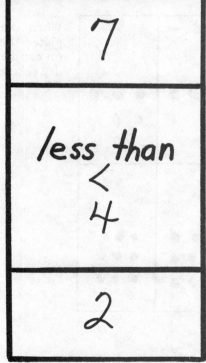

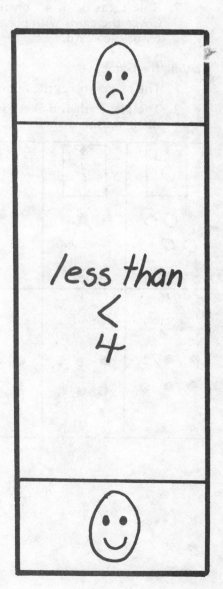

50

IS IT MORE OR LESS?

Objective

To make sets equivalent to a given numeral that have more or less than the matching number of elements

Materials

1. Bristol board, 22" x 28"
2. Folding bristol
3. Ten drapery hooks
4. Marking pens, crayons, ruler, paper cutter, paper punch
5. Clear adhesive plastic

Making the Game

1. Rule the bristol board into ten equal sections. Draw sets of dots of from one to ten in numerical order in the sections. Cover the board with plastic. Insert a drapery hook in the top center of each section.
2. Rule cards 3" x 4" on the folding bristol. Write the numerals one to ten on the cards. Cover the cards with plastic. Cut apart with the paper cutter. Punch a hole in the top center of each card.

Playing the Game

1. The player mixes the numeral cards and places them on the hooks in random order.
2. The player then makes each set match the numeral by either crossing out dots or drawing them in using a crayon.

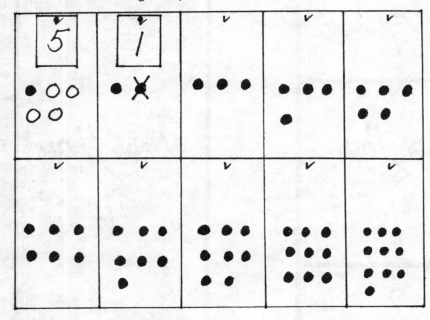

When the player has completed the activity, the crayon can be rubbed off with a dry tissue or cloth.

MORE OR LESS BINGO

Objective

To determine which of two numerals is more and which is less

Materials

1. Folding bristol (1 sheet each of green and red, several white)
2. Marking pens, ruler, paper cutter
3. Clear adhesive plastic

Making the Game

1. Make bingo cards by ruling the white bristol into 6" x 12" sections. Divide each section into two rows of four spaces each. Randomly write the numerals from one to ten in the spaces. Make sure that no two rows have the same set of numbers. Cover the cards with plastic. Cut into individual cards.
2. On the green paper, make stimulus cards from zero to nine. On the red paper, make stimulus cards from one to ten. Cover with plastic and cut apart.

Playing the Game

When teaching "more," use the green stimulus cards. When teaching "less," use the red cards.

1. The leader chooses a green card and calls the number.
2. The players cover a number on their bingo cards that is more than the number called.
3. The first player to cover all the numbers in a row wins the game.

For a higher skill activity, mix the red and green cards, requiring the players to cover both "more" and "less" numbers.

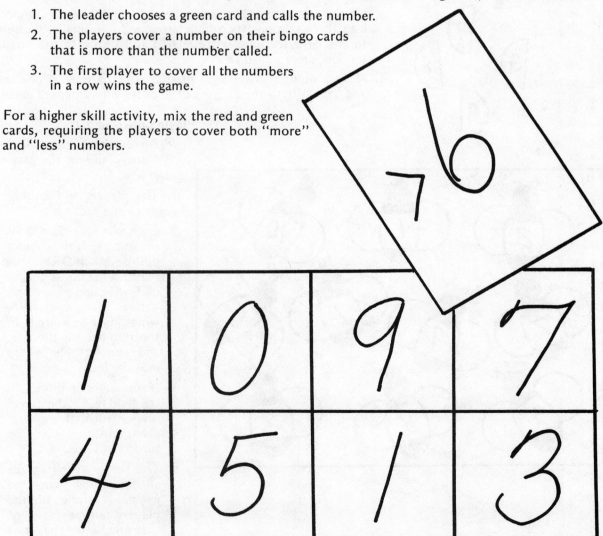

BEE BUZZ

Objective

To determine which of two numerals is more and which is less

Materials

1. Bristol board, 18" x 24" and 8" x 8"
2. Folding bristol
3. Marking pens, ruler, scissors, paper punch, paper fastener, compass
4. Clear adhesive plastic

Making the Game

1. On the large bristol board, draw ten 4" oval bee bodies with 3" diameter half-circle wings. Use a black marking pen to decorate and outline the bees. On the bee bodies, randomly write the numbers one to ten. Cover the board with plastic.

2. On the folding bristol, make ten 3" circles. Divide each in half. On the left half, write a number from one to nine. On the right half, write a number two to eleven. Cover with plastic and cut each circle in half.

3. On the bristol board square, draw a 7" circle. Divide the circle into three sections (two large areas and one smaller section). In the large sections, write MORE and LESS. In the smaller section, draw a "sad face."

4. Make a pointer, cover with plastic, punch a hole at one end, and attach it to the center of the circle with a paper fastener.

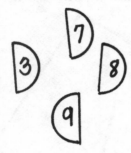

Playing the Game

1. Divide the twenty wings equally among the players.

2. The players will arrange the wings face up.

3. Each player in turn spins the spinner. If the spinner points to MORE, the player will choose a right half-wing and place it on a bee. (The right half-wings must be more than the number on the bee, and the left half-wings must be less.)

4. If the spinner points to the "sad face," the player loses a turn. If a player cannot make a play, his turn is forfeited.

5. The first player to place all his wings wins the game. If none of the players can make a play, the one with the least number of wings left is the winner.

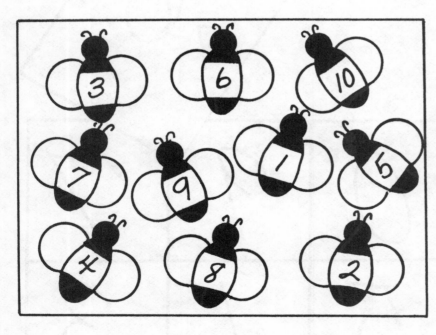

WHO HAS MORE?

Objective

To determine which of two numbers is more and which is less

Materials

1. Folding bristol
2. Marking pens, ruler, paper cutter
3. Forty poker chips
4. Clear adhesive plastic

Making the Game

1. Rule the folding bristol into forty 3" x 4" cards. Number the cards from one to ten, making four sets. On each card, draw a number of dots matching the numeral.
2. Cover with plastic and cut out.

Playing the Game

This is an activity for two players.

1. Deal out all the cards. Each player keeps his cards face down in a pile.
2. Each player takes a top card from his pile and places it face up. The player who has the "more" number wins the round, and the player who has the "less" number must give the other player a number of poker chips equal to the difference of the two numbers on the cards.
3. Play continues until one player wins all the chips. If each player has some chips after all the cards are played, the one who has the most chips is the winner.

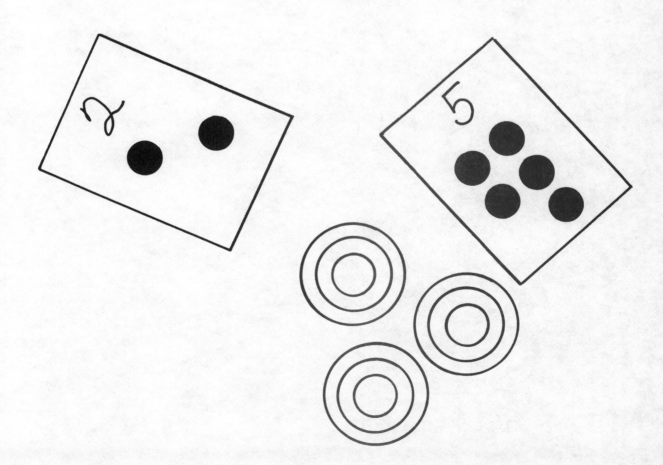

Numbers

LACE TO LACE

Objective

To recognize and name the numbers one to twelve

Materials

1. Bristol board, 8" x 16"
2. Marking pen, paper punch
3. Shoelaces
4. Clear adhesive plastic

Making the Game

1. On the left side of the bristol, mark numbers in random order from one to twelve, and again on the right side of the board.
2. Cover the board with plastic.
3. Next to each number, punch a hole. Put a lace through each hole on the left side, tying a knot on the back.

Playing the Game

1. The player names the numeral at the top on the left side.
2. He finds the match on the right side and puts the lace through the hole next to the match.
3. Play continues until all numbers are named and matched.

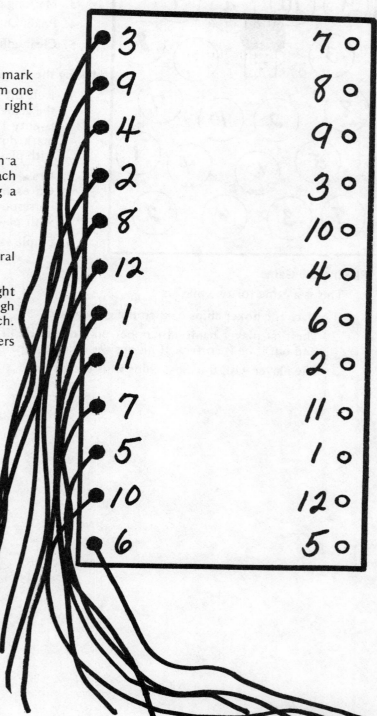

WHAT'S UNDER THE CIRCLE?

Objective

To recognize and name the numbers one to ten

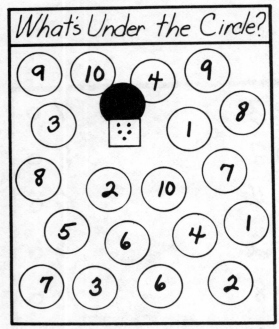

Materials

1. Bristol board, 10" x 14"
2. Folding bristol
3. Marking pen, scissors, stapler, ruler
4. Poker chips
5. Clear adhesive plastic

Making the Game

1. Rule off 1½" at the top of the board and label the game "What's Under the Circle?". Rule off twenty 1" square boxes. In each box randomly mark dots from one to ten. Cover the board with plastic.
2. Draw twenty 1½" circles from folding bristol. On each circle, mark numbers one to ten to correspond to the dots in the boxes. Cover with plastic and cut out the individual circles.
3. Staple each circle over the corresponding dots.

Playing the Game

This is a game for two players.

1. Place the poker chips next to the gameboard.
2. The first player names a number on a circle. He lifts up the circle and counts the dots to find out if he is correct. If his answer is correct, he takes one chip.
3. The player with the most chips wins the game.

NUMBERS, NUMBERS, NUMBERS

Objective

To recognize and name the numbers one to nine

Materials

1. Bristol board
2. Paper fasteners, paper punch, scissors, marking pen
3. Clear adhesive plastic

Making the Game

1. Draw numbers one to nine, 3" high.
2. Make as many numbers as the number indicates (for example, six of number six).
3. Cover with clear plastic.
4. Cut out each numeral. Fasten all numbers that are the same with a paper fastener.
5. Punch as many holes in each number as that number indicates.

Playing the Game

1. The child looks at each set of numerals and names them.
2. If the child does not know the name of the numeral, he can count the number in the set or count the number of holes in each numeral.
3. The child puts the numbers in numerical order.

For a variation, all numerals can be separated and the child can match all the similar numbers.

NUMBER NEWS FIND

Objective

To recognize the numbers one to ten

Materials

1. Newspaper pages showing supermarket ads
2. Colored marking pens

Playing the Game

1. Each child takes a newspaper page.
2. The teacher names a number.
3. Using the colored marking pens, the children find and circle the number wherever it appears on the newspaper page.
4. The player with the most circles on his page wins the game.

TOSS AND MATCH

Objective

To match dots to numbers

Materials

1. Two blank dice or styrofoam cubes
2. Marking pen
3. Score sheets

Making the Game

1. On one blank die, mark numbers from one to six. (If the pen marks rub off the die, apply adhesive tape to each side and then mark the tape.)
2. On the other blank die, mark from one to six dots.
3. On the score sheets write Player 1 and Player 2 on the top.

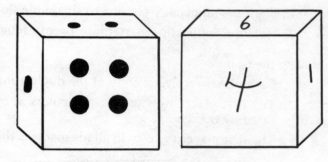

Playing the Game

1. Each child takes a turn in tossing the dice. If the dots and numerals match, the player marks a point on his section of the score sheet and takes another turn.
2. The play continues with the players taking turns and marking the score sheets if they get a point.
3. The child with the most points is the winner.

COLOR THE STRIP

Objective

    To recognize and name the numbers one to six

Materials

1. Duplicating masters
2. Ruler, pencil, crayons
3. Styrofoam cube

Making the Game

1. Mark the numerals from one to six on the sides of the foam cube.
2. Rule the duplicating master into twenty squares.

Playing the Game

1. Each player receives a copy of the duplicating master and a box of crayons.
2. The first player rolls the die. He colors as many spaces on the duplicating master as indicated by the die.
3. The first player to color in all his spaces is the winner.

The play may be extended by requiring an exact roll of the die in order to color in the last spaces.

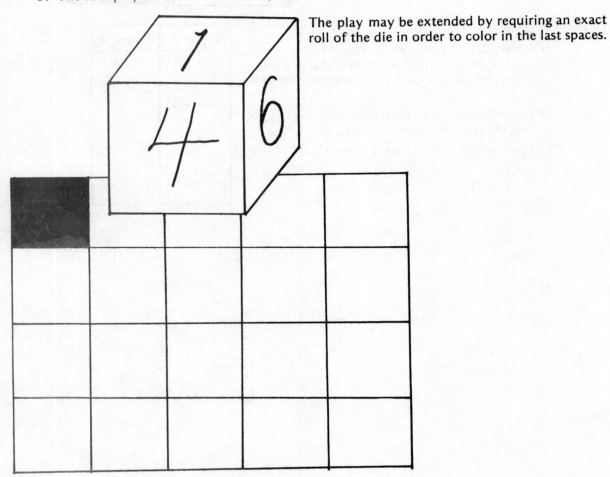

NUMBER BINGO

Objective

    To recognize and name the numbers zero to ten

Materials

1. Folding bristol
2. Marking pens, ruler, paper cutter
3. Colored chips
4. Clear adhesive plastic

Making the Game

1. Rule the folding bristol into 5" x 12" cards. Divide each card into two rows of four spaces each. Randomly write the numerals zero to ten in the spaces. Make sure that no two rows contain the same sets of numerals. Cover the cards with plastic.
2. Make a set of stimulus cards numbered zero to ten. Cover the cards with plastic and cut apart.

Playing the Game

1. Each player is given a bingo card.
2. Shuffle the stimulus cards. Place them face down on the playing area. Turn over the top card, name it, and show it to the group.
3. The players who have a match on their cards cover the space with a chip.
4. The first player to get all the numerals in a row wins the game.

To vary the activity, make cards that contain sets of dots as well as numerals. An empty space indicates zero.

For a higher skill, name the numeral but don't show it.

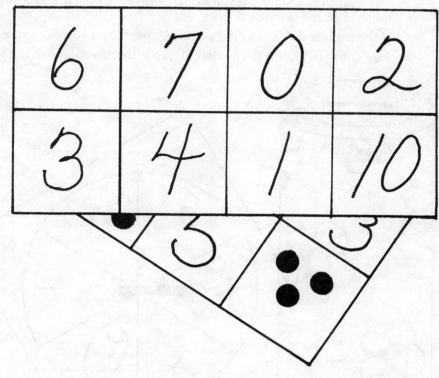

GET TO ZERO

Objective

To understand the concept of zero

Materials

1. Bristol board, 6" x 6"
2. Colored chips
3. Empty cylindrical potato chip can with plastic cover
4. Marking pens, paper punch, paper fastener
5. Clear adhesive plastic

Making the Game

1. Decorate the can. Make a slit in the plastic top.
2. Draw a 5" circle on the bristol square. Divide the circle into six sections. Write the numbers one to five in the sections and draw a "sad face" in the sixth section. Cover the board with plastic.
3. Make a pointer, cover with plastic, punch a hole in the end, and use the paper fastener to attach it to the center of the circle.

Playing the Game

1. Discuss zero with the group. Tell them that the object of this game is to be the first player to have zero chips.
2. Give each player twenty chips.
3. The first player spins the spinner. He drops into the can as many chips as indicated by the number on the spinner.
4. If the spinner points to the "sad face," the player forfeits a turn.
5. Play continues in turn. The first player with zero chips wins.

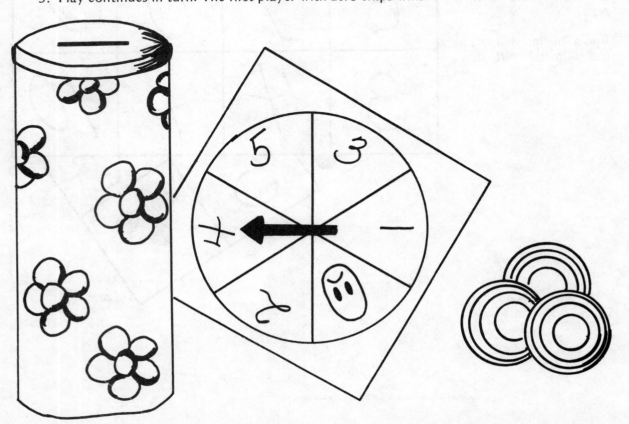

NUMBER PUZZLE MATCH

Objective

To match sets to numerals

Materials

1. Bristol board, 15" x 18"
2. Stickers
3. Marking pen, ruler, scissors
4. Clear adhesive plastic

Making the Game

1. Rule the bristol board into ten 3" x 9" rectangles. In the top half of each rectangle, mark numerals from one to ten. On the bottom half, place from one to ten stickers to correspond to the numeral.
2. On the reverse side, repeat the numeral. Below it, write the numeral name.
3. Cover with plastic. Cut out each rectangle, then cut each one into an interesting puzzle shape.

Playing the Game

1. Separate the cards into two piles, with all the top pieces in one pile and all the bottom pieces in the other.
2. The player selects a top puzzle piece. He names the numeral it shows.
3. He then looks in the second pile and finds the piece having the matching number of stickers. If the puzzle pieces fit, the player is correct.
4. Play continues until all pieces are matched.

For a more difficult skill, reverse the bottom portion. The child will match the numeral to the number name.

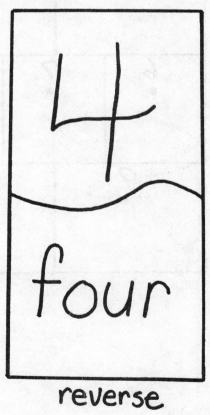

reverse

THE COUNTING BOARD

Objective

To count the number of elements in a set and match it to the correct numeral

Materials

1. Bristol board, 12" x 18"
2. Folding bristol
3. Marking pens, ruler, paper cutter, paper punch
4. Ten small paper fasteners
5. Set of pictures of from one to ten objects
6. Clear adhesive plastic

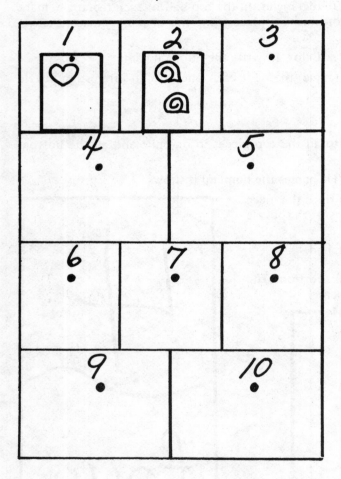

Making the Game

1. Divide the bristol into ten sections. Number the sections one to ten. Cover the board with plastic. Punch a hole in the top of each section and put a paper fastener through each hole.

2. Divide the folding bristol into 3" x 4" cards. Place from one to ten pictures on each card. To make the activity self-correcting, write the numeral on the back of each card. Make several sets of cards.

3. Cover with plastic and cut apart with the paper cutter. Punch a hole in the top of each card.

Playing the Game

1. Give the child several sets of picture cards.

2. He will count each set and hang it in the correct section of the board.

I'VE GOT YOUR NUMBER

Objective

To match numbers to sets of objects

Materials

1. Bristol board, two sheets
2. Stickers
3. Marking pen, scissors
4. Clear adhesive plastic

Making the Game

1. Draw five telephones down each side of a sheet of bristol board. On each telephone, place sets of from one to ten stickers. Cover the gameboard with plastic.
2. Draw ten telephones on the other sheet of bristol board. Number the phones from one to ten. Cover with plastic. Cut out the individual phones.

Playing the Game

1. The child picks up an individual phone and names the number marked on it.
2. He finds a telephone on the gameboard that corresponds with the number on the individual phone. He places the individual phone over the matching telephone on the gameboard.
3. Play continues until all phones are matched.

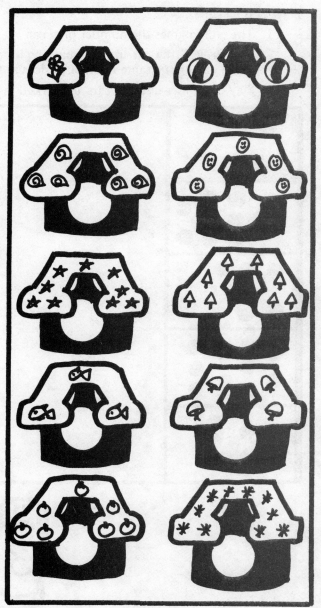

PARKING LOT

Objective

To match numbers one to ten with corresponding dots

Materials

1. Bristol board, 9" x 12"
2. Marking pen, adhesive tape, ruler
3. Ten small plastic cars
4. Clear adhesive plastic

Making the Game

1. Rule the bristol board into five equal rectangles down each side. Mark from one to ten dots randomly in each rectangle. Cover the board with plastic.
2. Put a piece of adhesive tape on each car. Marking on the tape, number the cars from one to ten.

Playing the Game

1. The child names the number on a car.
2. He "drives" the car into the parking lot and parks the car in the space reserved for it (the space having the same number of dots as the number on the car).
3. Play continues until all cars are parked in all spaces.

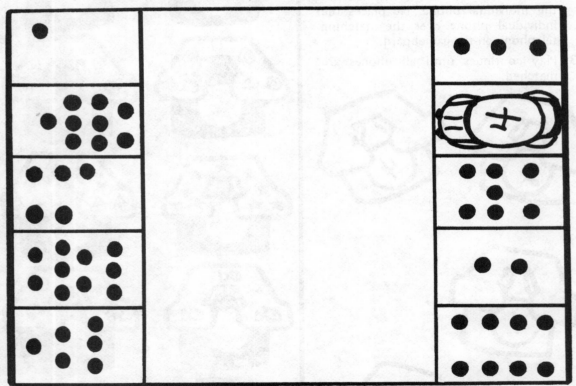

PIN THE NUMBER

Objective

To match numerals to corresponding number of dots

Materials

1. A cardboard circle (for example, a pizza board)
2. Marking pen, ruler
3. Ten spring–type clothespins
4. Clear adhesive plastic

Making the Game

1. Divide the circle into ten equal pie–shaped wedges. Mark dots from one to ten in each wedge. Cover the circle board with plastic.
2. Mark a numeral from one to ten on the head of each clothespin.

Playing the Game

1. The child names the number on each clothespin.
2. He finds the same number of dots on the pie–shaped circle and attaches the clothespin to its match.
3. When all clothespins are used, the activity is completed.

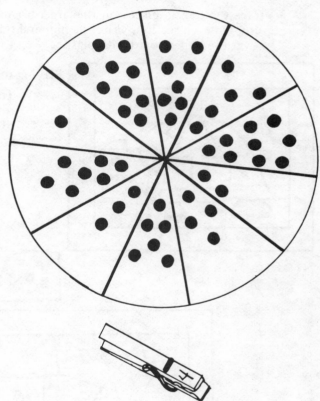

FIT THE FORM

Objective

To match dots to numerals

Materials

1. A large, colorful picture
2. Two pieces of bristol board, one the same size as the picture, and one larger
3. Marking pen, paste, scissors
4. Clear adhesive plastic

Making the Game

1. Paste the picture on the piece of bristol board of the same size. Cover with plastic. Cut into ten interesting puzzle shapes.
2. On the reverse side of each piece, make dots from one to ten.
3. Trace the puzzle pieces on the larger bristol board, fitting each piece to complete the picture. On each section, write the numeral that corresponds to the dots on the puzzle piece. Cover the larger board with plastic.

Playing the Game

1. The child takes each piece of the puzzle, reads the number of dots on the reverse side of the picture, and matches the form to its corresponding number on the puzzle board.
2. Play continues until the puzzle is completed.

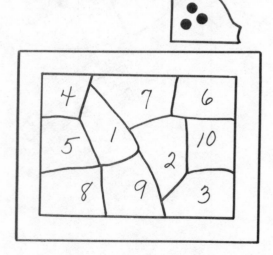

A DOZEN EGGS

Objective

To recognize and count from one to twelve

Materials

1. An empty egg carton
2. Twelve plastic dividing egg shapes
3. Marking pen
4. Seventy-eight beans

Making the Game

1. Mark each egg with a numeral from one to twelve.

Playing the Game

The child names the numeral on each egg and puts the corresponding number of beans inside each egg. Play continues until all the beans are inside all the eggs.

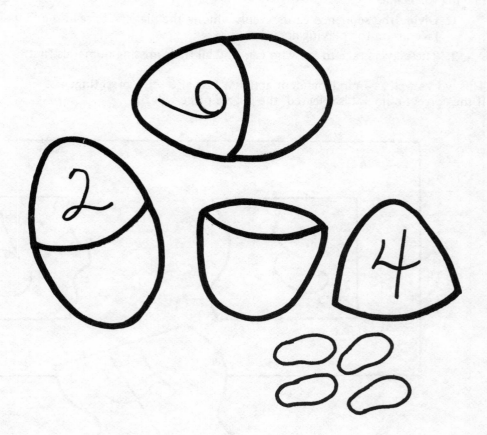

THAT'S FITTING

Objective

To name the numeral that comes before or after a given numeral

Materials

1. Two contrasting sheets of bristol board
2. Marking pens, ruler, scissors, paper cutter, paste
3. Clear adhesive plastic

Making the Game

1. Rule the bristol into 3" x 9" cards. Write three numerals in sequence on the cards. Cut one of the numerals off the card, using puzzle lines.
2. Rule the second sheet of bristol into 4" x 10" cards. Paste the sequence card (minus the "missing" number) onto the larger card. Cover these cards and the "missing" numeral card with plastic.

Playing the Game

1. Divide the sequence cards evenly among the players. Spread the "missing" numeral cards face up on the playing area.
2. The players race to see who can find all their missing numerals first.

This works well as an independent activity. It is also self-correcting:
if the correct numeral is selected, the puzzle piece will fit.

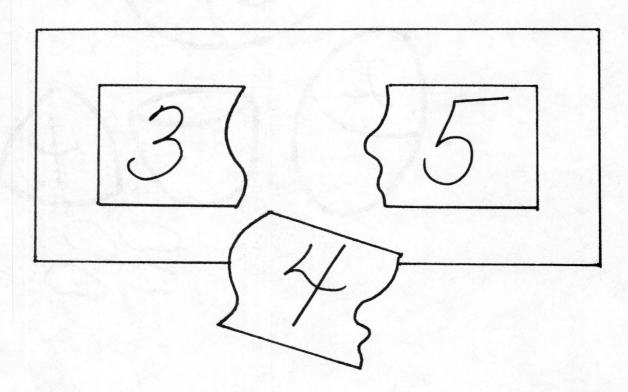

FLIP AND FIND

Objective

To identify missing numerals

Materials

1. Bristol board
2. Marking pen, ruler, scissors, paper punch, crayon, pipe cleaners

Making the Game

1. Rule the bristol board into six 4" x 8½" rectangles.
2. On one rectangle (which will be the cover page) mark in a number line, using dots and numerals from zero through ten.
3. On another rectangle, mark a section of a number line. Omit one number, drawing a blank square instead (see illustration).
4. Mark the remaining rectangles with number lines as above.
5. Cover with plastic, cut apart into cards, and punch two holes on the top of each card.
6. Cut two pieces of pipe cleaner and put through the holes to make a book. Twist the ends of the pipe cleaner to close.

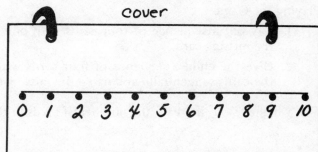

Playing the Game

1. The child names the numbers on the number line on each page.
2. He fills in the blank on the number line by drawing a line with crayon to the number that belongs in the blank square.
3. Play continues until all blanks are connected to the correct numbers.

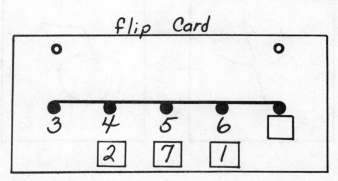

When the child has completed the activity, the crayon marks can be rubbed off with a dry tissue or cloth.

WHAT NUMBER IS MISSING?

Objective

To determine which numeral in a sequence is missing

Materials

1. Folding bristol
2. Marking pens, ruler, paper cutter
3. Clear adhesive plastic

Making the Game

1. Rule the paper into twenty-one 3" x 3" sections. Write the numbers zero to twenty on the sections.
2. Cover with plastic. Cut into cards with the paper cutter.

Playing the Game

1. Lay out a sequence of four cards with one numeral card missing. The player must name the missing card.
2. Give the child a sequence of four cards with one card missing. The player must lay out the cards sequentially and name the missing card.

For a higher skill, increase the number of cards in the sequence and the number of missing cards.

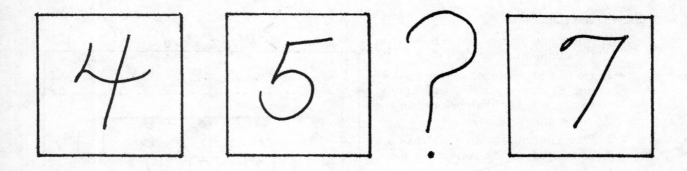

THE CATERPILLAR

Objective

> To name the numerals that are missing in a given sequence

Materials

1. Bristol board, 12" x 18"
2. Poker chip
3. Marking pens, crayon
4. Clear adhesive plastic

Making the Game

1. Following the illustration, make the "caterpillar" by tracing twenty-six circles, using the poker chip.
2. Decorate the caterpillar. Cover the board with plastic.

Playing the Game

1. Write the sequence of numbers on the caterpillar, leaving out several numbers.
2. The child will write in the missing numerals with a crayon. (Rub off the crayon marks with a dry cloth or paper towel.)

Use the caterpillar to reinforce counting by twos, fives, tens, even and odd numbers.

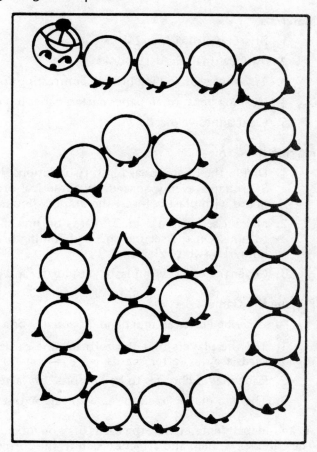

WHAT COMES NEXT?

Objective

To name the numeral that comes before or after a given numeral

Materials

1. Bristol board, 18" x 24"
2. Twenty small drapery hooks
3. Two sheets of folding bristol, contrasting colors
4. Marking pens, ruler, paper cutter, paper punch
5. Clear adhesive plastic

Making the Game

1. Divide the bristol board into two sections. In each section draw a vertical row of five 3" x 3" squares evenly spaced. In numerical order, write the numerals one to ten. Cover the board with plastic. Insert the drapery hooks on either side of the numeral squares.
2. Make a set of ten 3" x 3" cards on one sheet of folding bristol. Write the numerals zero to nine. On the second sheet of folding bristol, make another set of cards and write the numerals two to eleven.
3. Cover the cards with plastic, cut apart, and punch a hole in the top center of each card.

Playing the Game

This may be either an independent activity or a game for two players.

1. Tell the players that the zero to nine cards are the "before" cards and the two to eleven cards are the "after" cards.
2. Each player chooses to be "before" or "after."
3. The two players race to see who can be the first to hang up all their cards.

As an independent activity, the cards may be mixed.
The child sorts them and attaches each to the correct holder.

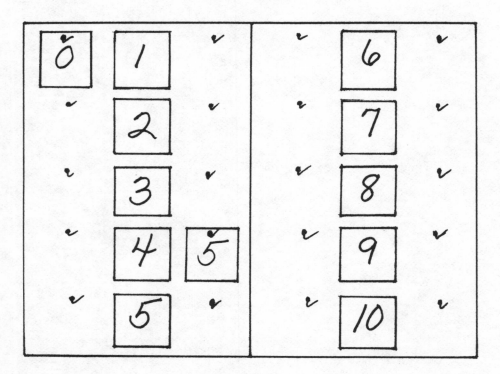

BLOCK THE CHIP

Objective

To identify numbers that come before or after a specific number

Materials

1. Bristol board, 12" x 18"
2. Marking pen, ruler, compass, paper fasteners
3. Sets of colored chips or buttons
4. Clear adhesive plastic

Making the Game

1. On the upper portion of the bristol, rule off sixty-four 1" squares, eight across and eight down. In each square, randomly mark numbers from zero to eleven.
2. On the bottom portion of the bristol, draw two equal circles. Divide each circle into eight equal sections. In each section, write a number from one to ten. Label one circle BEFORE and the other AFTER. Cover the board with plastic.
3. Make two pointers, cover with plastic, punch a hole at the end of each, and use a paper fastener to attach one to the center of each circle.

Playing the Game

The object of the game is for each player to cover as many numbers as he can in a row, diagonally, vertically or horizontally. The other players will try to block their opponents.

1. Each player chooses a set of colored chips.
2. The player chooses to spin either the BEFORE or the AFTER spinner. Using a chip, he covers the number on the gameboard that comes before or after the number indicated by the spinner.
3. The next player takes a turn, and play continues.
4. The winner is the player who has the most numbers covered in a row, diagonally, vertically or horizontally.

7	6	1	5	6	3	4	11
9	11	8	4	2	1	9	3
2	7	3	9	8	10	8	2
10	2	5	0	10	3	1	7
8	11	9	5	1	2	6	0
5	7	4	3	2	6	4	3
1	0	4	10	5	4	11	9
8	11	6	7	5	7	10	6

Before   After

Joining Sets

MATCH THE SUM

Objective

To match number problems with correct sums

Materials

1. Bristol board
2. Marking pens, ruler, scissors
3. Clear adhesive plastic

Making the Game

1. Draw a number of 3" x 9" rectangles on the bristol board. On the top portion of each rectangle, write an addition problem. On the bottom portion, write the sum of the problem.
2. Cover with adhesive. Cut apart into cards. Cut the bottom portion from the top with interesting puzzle lines.

Playing the Game

The player finds the correct sum for each number problem by matching the top portion of each rectangle with the bottom. If the puzzle piece fits, the player is correct.

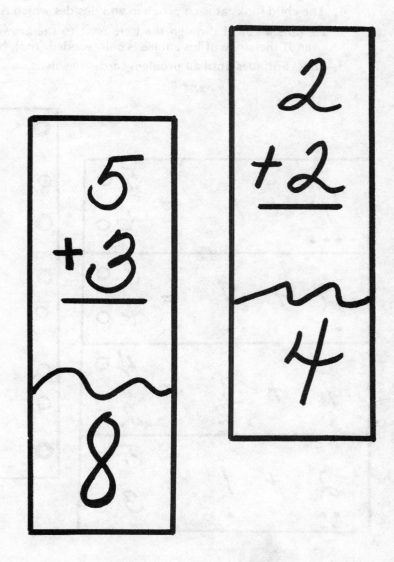

MAKE A CHOICE

Objective

To select the sum of two numbers

Materials

1. Several sheets of bristol board, 9" x 12"
2. Marking pens (black and red), pencil, ruler, paper punch
3. Clear adhesive plastic

Making the Game

1. Divide the bristol board sheets into 3" x 9" rectangles.
2. Using the black pen, write addition problems on the left side of the rectangles. Draw dots under the numbers to give clues.
3. On the right side of the rectangle, write two sums. One sum must be the correct answer.
4. Cover the sheets with plastic. Cut apart into individual cards.
5. Punch a hole next to each sum. Reverse the cards. Make a red circle around the hole next to each correct sum.

Playing the Game

1. The child looks at each problem and decides which is the correct sum.
2. He puts a pencil through the hole next to the answer he has chosen, and turns the card over. If the circle of his choice is color-coded (red), he is correct.
3. Play continues until all problem cards are solved.

CONE TOSS

Objective

To complete the operation indicated by –3, –2 and –1

Materials

1. Folding bristol, 12" x 18"
2. Marking pens, stapler, ruler, duplicating master
3. Table tennis ball
4. Clear adhesive plastic

Making the Game

1. Make six paper cones from the folding bristol. On three cones write –1. On two cones write –2. On one cone write –3. Cover the cones with plastic.
2. Rule the duplicating master into a column of ten 1" squares. Starting at the top, write the numerals ten to one.

Playing the Game

1. Staple the cones to the bulletin board, following the illustration (inverted triangle with –3 on the bottom).
2. Give each player a duplicated copy of the column with the numerals ten to one.
3. The first player throws the table tennis ball. Starting with square 10, he crosses out the number of squares indicated by the cone the ball landed in (–2 = cross out two squares). The first player to cross out square 1 is the winner.

For more fine motor work, have the player cut off the squares with scissors instead of crossing them out.

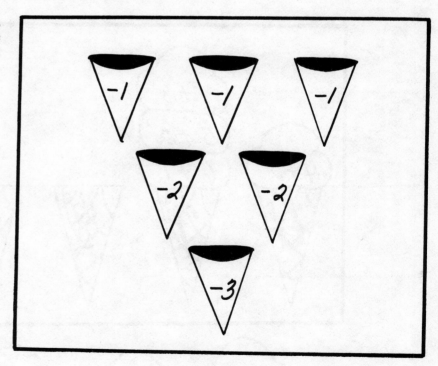

ICE CREAM CONE FIND

Objective

To solve addition problems with sums six through nine

Materials

1. Two sheets of folding bristol, 12" x 18"
2. Marking pen, scissors
3. Clear adhesive plastic

Making the Game

1. On one sheet of folding bristol, draw four ice cream cones. Mark one number on each cone.
2. On the other sheet of bristol, draw at least twelve scoops of ice cream. On each scoop, write addition facts that add up to the numbers (sums) on the cones (three facts for each cone).
3. Cover both sheets with plastic. Cut out the scoops.

Playing the Game

1. The player names the numerals on the cones.
2. He reads the addition problems on the scoops and places the scoop on the cone with the correct sum to solve the problem.
3. At the end of the activity, the player should have a triple ice cream cone.

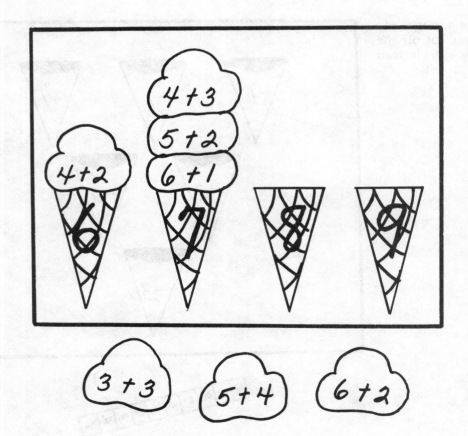

THE APPLE TREE

Objective

> To add sums of five to ten without manipulation

Materials

1. Bristol board, 22" x 28" and 8" x 8"
2. Folding bristol (brown, green, red)
3. Small drapery hooks
4. Marking pens, paste, scissors, paper punch, paper fastener, compass, ruler
5. Clear adhesive plastic

Making the Game

1. Make an apple tree from the green and brown folding bristol. Cut it out and paste it to the large bristol board. Cover with plastic. Place drapery hooks on the tree.
2. Draw a 6" circle on the bristol board square. Divide it into six equal sections. Mark the numbers five to ten in the sections. Cover with plastic.
3. Make a pointer, cover it with plastic, cut it out, and attach it to the center of the circle with a paper fastener.

4. Make fifty-one 2" diameter apples on the red folding bristol. On the apples, write addition problems from the "families" five through ten. (For example, the "five family" has six possible combinations: 5 + 0; 4 + 1; 3 + 2; 2 + 3; 1 + 4; 0 + 5.)
5. Cover with plastic. Cut out the individual apples. Punch a hole in the top of each apple.

Playing the Game

1. Divide the apples equally among the players, who place them face up.
2. The first player spins the spinner. He hangs up one of his apples having the sum indicated on the spinner (for example, if the spinner points to five, he will hang up an apple from the "five family").
3. Play continues until one player wins the game by hanging all his apples on the tree.

To make the game last longer, place a "sad face" on the spinner. If the spinner points to the "sad face," the player forfeits a turn.

This activity can be expanded by using three-digit "families."

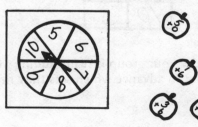

TOSS AND ADD

Objective

To add two sets of from zero to five objects

Materials

1. Bristol board, 22" x 28"
2. Folding bristol
3. Pictures of sets of from one to five objects (two of each)
4. Marking pens, ruler, pencil, paper cutter
5. Two foam cubes, beanbags or blocks
6. Sets of ten chips or buttons (one set for each player)
7. Clear adhesive plastic

Making the Game

1. Rule the bristol board into three rows of four spaces. Randomly place the sets of pictures in the spaces, leaving two spaces blank (for zero). Cover the board with plastic.

2. Rule the folding bristol into 6" x 9" sections. Divide each section into ten spaces. Randomly place the numbers one to ten in the spaces. Cover with plastic. Cut apart into individual scorecards.

Playing the Game

1. Place the gameboard on the floor. Give each player a scorecard and a set of chips.

2. The first player tosses the cubes onto the board. He adds up the two numbers the cubes land on. He covers the number on his scorecard that matches the sum of the numbers on the board.

3. The first player to cover all the numbers on his scorecard wins the game.

You know your group is really adding when you see them figure out in advance which sets they need to win!

FLOWER POWER

Objective

To identify sums of two numbers

Materials

1. Bristol board, 12" x 18"
2. Marking pens, paper fastener
3. One playing piece for each player
4. Clear adhesive plastic

Making the Game

1. Draw a flower path around the outer edges of the bristol board. On each flower, write an addition problem.
2. In the center of the gameboard, draw ten petals in the shape of a circle. On each petal write a sum that matches the problems on the flowers. Use sums one to ten.
3. Draw arrows at the top left corner of the board, showing start and finish points.
4. Cover the board with plastic.
5. Make a pointer, cover it with plastic, punch a hole in the end, and use a paper fastener to attach it to the center of the petal circle.

Playing the Game

1. The first player takes a turn by spinning the spinner. He places his playing piece on the first flower along the path that has the sum indicated on the spinner.
2. Play continues with the next player taking his turn.
3. The first player to reach the finish point wins the game.

FORWARD OR BACK?

Objective

To solve addition and subtraction problems

Materials

1. Bristol board, 12" x 18"
2. Folding bristol
3. Marking pens, ruler, paper cutter
4. One playing piece for each player
5. Clear adhesive plastic

Making the Game

1. Make a track of 2" squares around the bristol board. Decorate it with a picture if you wish (we have drawn a dinosaur). Cover the gameboard with plastic.
2. Rule the folding bristol into thirty 4" x 4" sections. In each section write addition and subtraction problems with sums from one to ten. Cover with plastic and cut apart into cards.

Playing the Game

1. Put the cards face down in a pile in the middle of the gameboard.
2. Each player puts his marker on Start.

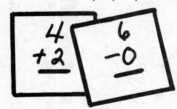

3. The players take turns drawing the top card from the pile. If the problem is an addition problem, the player gives the correct sum and moves his marker forward that number of spaces. If it is a subtraction problem, the player gives the correct answer and moves backward that number of squares that equals the answer.
4. The first player to reach the finish box wins the game.

WHAT'S THE DIFFERENCE?

Objective

To identify differences of subtraction problems from zero to five

Materials

1. Bristol board
2. Marking pens, paper cutter
3. Clear adhesive plastic

Making the Game

1. Draw twenty-one 3" x 6" rectangles on the bristol board. On each rectangle write subtraction problems having the differences from zero to five.
2. Cover with plastic and cut apart into individual cards.

Playing the Game

1. Place all the cards face down on the table.
2. The first player calls out a number from one to five. He then turns over the top card. If the answer to the subtraction problem is the same as the number called, the player keeps the card. If the answer is not the same, he replaces the card on the bottom of the deck.
3. Play continues in turn until all cards are taken.
4. The player with the most cards wins the game.

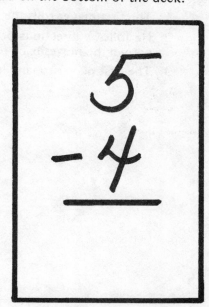

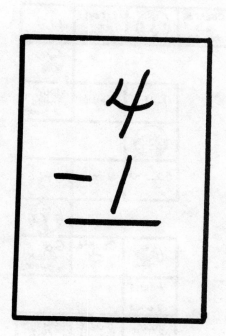

SUPER MATH BOARD

Objective
To reinforce the development of number skills

Materials
1. Bristol board, 12" x 18"
2. Marking pens
3. Stickers
4. A die
5. Clear adhesive plastic

Making the Game
1. Make a gameboard with 1" x 1½" rectangles. In each box write math problems to match the progress of the class. Vary the problems and include instructions (for example: count by twos to twenty, tell the time indicated on a clock picture, miss a turn or move ahead several spaces). Use stickers to indicate "free" spaces.
2. Cover the board with plastic.

Playing the Game
1. The first player rolls the die and moves that number of spaces on the gameboard.
2. He follows directions or gives the answer to the problem in the box. If the answer is not correct, he moves back to the starting point.
3. The first player to land in the finish box wins the game.

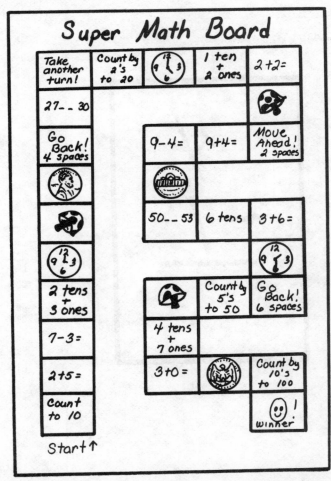

MATCH THE SIGNS

Objective

To recognize and match signs

Materials

1. Blank dice or two small blocks
2. Marking pen, adhesive tape
3. Scoring sheets

Making the Game

1. Place a piece of adhesive on each side of the dice.
2. On the adhesive, mark the following signs: $+, -, =, \cancel{c}, <, >$.

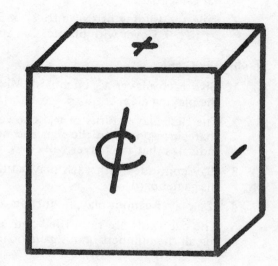

Playing the Game

1. Give each player a blank sheet of paper for a scoring sheet.
2. The first player tosses the dice. If the signs match, the player marks a point on his score sheet and takes another turn.
3. Play continues with each player taking a turn and marking his score.
4. The player with the highest score at the end of game time is the winner.

FOLLOW THE SIGN

Objective

To recognize and identify the signs < and >

Materials

1. Bristol board
2. Folding bristol
3. Marking pens, paper cutter
4. Ten markers per player
5. Clear adhesive plastic

Making the Game

1. Rule the bristol board into 6" x 12" cards. Make two rows of five spaces on each card. In numerical order, write the numerals one to ten. Cover with plastic. Cut apart into individual gameboards.
2. Rule the folding bristol into 2" x 3" cards. Write the signs and numerals <2 to <10, and >1 to >9. Cover with plastic. Cut apart into individual cue cards.

Playing the Game

1. Give each player a gameboard. Mix the small cue cards and place them in the center of the playing area.
2. The first player turns over a cue card. He places a marker on a number on his gameboard that corresponds to the sign and number on the cue card. (For example, a >5 cue card indicates that the player will cover any number that is greater than 5.)
3. Play continues with each player turning over a card and placing a marker on a number on his gameboard.
4. If a player cannot play, he forfeits a turn.
5. The cue cards are reshuffled and used again until one player has won the game by covering all the numbers on his gameboard.

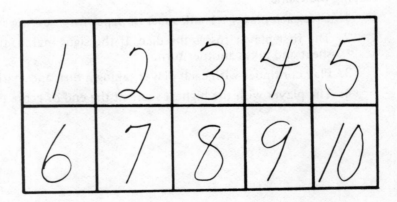

PLUS AND MINUS

Objective

To correctly read the signs + and –

Materials

1. Folding bristol, 12" x 18" and 6" x 6"
2. Twenty-four chips
3. Marking pens, ruler, compass, paper punch, paper fastener
4. Seals or stickers
5. Clear adhesive plastic

Making the Game

1. Divide the large board into two sections. In each section, make a track having twelve circles. (Use a chip to trace the circles.) Decorate the board by randomly placing a few seals.

2. On the square piece of bristol, make a 5" circle. Divide it into eight equal sections. Randomly place the following signs and numerals in the sections: +1, +2, +3, +4, –1, –2, –3, –4. Cover with plastic.

3. Make a pointer, punch a hole in the end, and attach it to the center of the circle with the paper fastener.

Playing the Game

This is a game for two players.

1. Each player takes twelve chips and chooses a side of the board.

2. The first player spins the spinner. If the pointer shows a plus number, he puts an equal number of chips on circles in his track. If the pointer shows a minus number, he takes an equal number of chips off the circles on his track.

3. Play continues with the second player taking his turn.

4. If a player spins a minus number and has no chips on his track, he cannot make a play and the turn is forfeited. (If a player spins –4 and has only three chips on the board, he must remove all three chips.)

5. The first player to cover all the circles on his track wins the game.

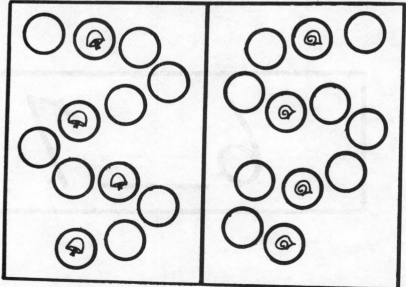

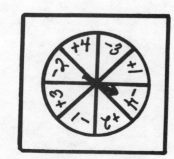

94

FILL IN SIGNS

Objective

To solve problems by filling blanks with correct signs

Materials

1. Bristol board
2. Marking pens, crayon
3. Clear adhesive plastic

Making the Game

1. Make several 4" x 12" cards.
2. On each card write a problem, omitting the signs that indicate whether it is a problem in addition or subtraction, or whether the sums are equal, greater or less.
3. Cover the cards with clear plastic.

Playing the Game

1. The child looks at each card and decides whether to use the signs +, –, =, < and >.
2. He fills in the blanks with a crayon.

When the child has completed the activity, the crayon marks can be rubbed off with a dry tissue or cloth.

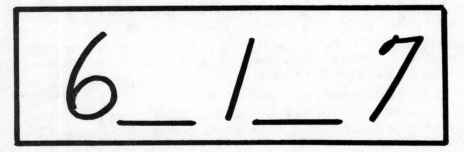

HOUSE MATCH

Objective

To match expanded numbers to double–digit numbers

Materials

1. Folding bristol, 12" x 18"
2. Marking pen, crayon
3. Clear adhesive plastic

Making the Game

1. Draw houses down the left side of the game-board. On the roof of each house, write an expanded number.
2. On the right side of the gameboard, draw doors opposite the houses. On each door, in random order, write a double–digit number to match the expanded numbers on the house.
3. Cover the board with clear adhesive plastic.

Playing the Game

1. The child looks at the house on the left side of the card.
2. He matches the numbers on the house with the door numbers by drawing a line with a crayon.

After all houses are matched, the crayon marks can be rubbed off with a dry tissue or cloth.

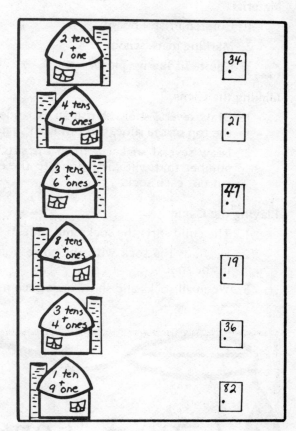

SOCK IT TO ME

Objective

To match expanded numbers to double-digit numbers

Materials

1. Colored bristol board
2. Marking pens, scissors
3. Colored adhesive plastic

Making the Game

1. Draw several shoes. Write an expanded number on each. Cover with plastic. Make a slit in the top of the shoe. Cut out each shape.
2. Draw several socks to fit in the slits of the shoes. On each sock, write the double-digit number that will correspond to the expanded numbers on a shoe. Cover with plastic and cut out each sock.

Playing the Game

1. The child sorts the socks and shoes.
2. He finds the sock with the corresponding double-digit number and tucks it into the slit in the shoe.
3. When all socks and shoes have been matched, the activity is completed.

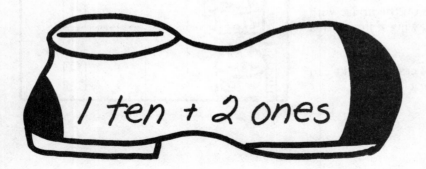

TENS TO ONES MATCH

Objective

To match expanded numbers to double–digit numbers

Materials

1. Folding bristol
2. Marking pen
3. Clear adhesive plastic

Making the Game

1. Rule the bristol into 2" squares and an equal number of 2" x 5" rectangles.
2. On each 2" square, write a double–digit number.
3. On each rectangle, write the expanded number to match the double–digit number.
4. Cover with clear adhesive plastic and cut apart into cards.

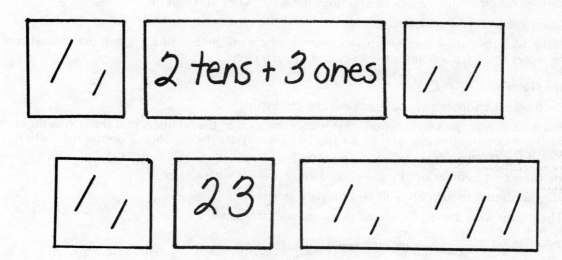

Playing the Game

1. Place all cards face down on the playing area.
2. The first player turns over two cards, one a square and the other a rectangle. If they match, the player takes them and gets another turn. If the cards do not match, the player replaces them face down.
3. Each player takes a turn and tries to recall where the matching cards are on the table.
4. The player with the most cards is the winner.

WHEEL A NUMBER

Objective

To identify double–digit numbers

Materials

1. Bristol board
2. Marking pens, ruler, compass, paper punch, paper fasteners
3. Sets of colored chips or buttons
4. Clear adhesive plastic

Making the Game

1. Mark an 8" x 9" rectangle at the top of the bristol board. Rule the rectangle into 1" sections. In each section write a double–digit number (eleven through eighty-nine).
2. On the bottom of the board, make two circles. Divide the left circle into eight pie–shaped wedges. In each section, mark the "tens" numbers from one through eighty. Divide the right circle into nine wedges and number the wedges one through nine.
3. Cover the board with plastic.
4. Make two pointers, cover with plastic, punch a hole on the end of each, and use the paper fasteners to attach a pointer to the center of each circle.

Playing the Game

1. Each player chooses a set of colored chips or buttons.
2. The first player spins both wheels. He names the sum of the numbers that the arrows point to (for example, the number 20 on the left wheel plus the number 2 on the right wheel equals the number 22).
3. He finds that number on the gameboard and covers it with a chip.
4. Play continues in turn until all numbers are covered.
5. The player with the most covered numbers wins the game.

BUILD A NUMBER

Objective

To develop the understanding of ones, tens, and hundreds

Materials

1. Bristol board, three different colors
2. Marking pens, ruler, scissors
3. Clear adhesive plastic

Making the Game

1. Rule off nine 3" x 3" cards of one color. Rule off nine 3" x 6" cards of a second color. Rule off nine 3" x 9" cards of a third color.
2. On the 3" x 3" cards, write the numerals one through nine. These are the ones cards.
3. Divide each 3" x 6" card into two sections. In each section write 1 / 0 to 9 / 0. These are the tens cards.
4. Divide each 3" x 9" card into three sections. In each section write 1 / 0 / 0 to 9 / 0 / 0. These are the hundreds cards.
5. Make an instruction card with a variety of one-, two-, and three-digit numbers.

Playing the Game

1. Give the child the instruction card.
2. He will duplicate each number on the card, using the ones, tens, and hundreds cards. The ones card will always be on the top at the right, the tens card will be in the middle, and the hundreds card will be on the bottom (see illustration).

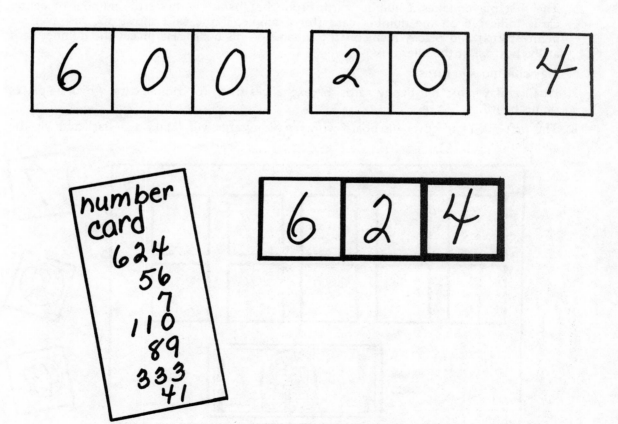

MAKE A DOLLAR

Objective

    To recognize that ten ones equal one ten, and ten tens equal one hundred

Materials

1. Bristol board
2. Coin stamps (thirty penny coin stamps, thirty dime coin stamps)
3. Play dollars
4. Clear adhesive plastic

Making the Game

    This is a game for four players.

1. Make four 9" x 12" boards. On each board draw ten 2" x 2" boxes (five in each row). Under the boxes, mark an equals sign and glue a play dollar.
2. Rule bristol board to make eighty 2" squares. Put a penny coin stamp on thirty squares. Put a dime coin stamp on thirty squares. On twenty squares, write the numbers one to ten.
3. Cover all with clear adhesive plastic. Cut apart into individual cards.

Playing the Game

1. Place all the number cards in a pile face down.
2. Put all penny cards together.
3. Put all dime cards together.
4. Each player chooses a gameboard.
5. The first player takes a number card. From the "bank" he takes the number of penny cards indicated on the number card (for example, if the card shows six, he takes six penny cards) and places them on the squares on his board. He places the number card at the bottom of the pile.
6. Play continues in turns.
7. When a player has ten penny cards, he exchanges them for a dime card, which he places on his board.
8. The first player to cover his board with ten dime cards will "Make a Dollar" and win the game.

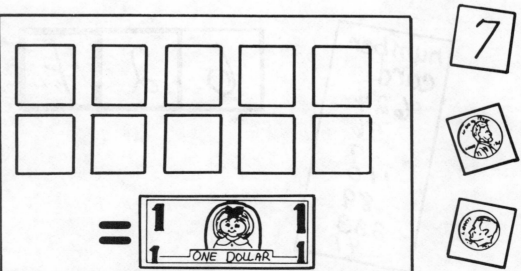

Measurement

LONGEST OR SHORTEST?

Objective

> To identify longest or shortest pictures

Materials

1. Bristol board, 9" x 12"
2. Marking pens, pencil, paper punch
3. Clear adhesive plastic

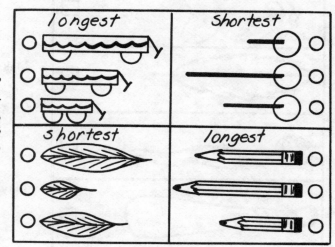

Making the Game

1. Divide the bristol board into 4" x 4½" boxes. In each box draw three pictures (one short, one middle–sized, and one longest). Label instructions on the top of each box (longest, shortest).
2. Next to each picture punch a hole. On the other side of the correct hole, color code with a red marking pen.
3. Cover with clear adhesive plastic.

Playing the Game

1. The child looks in each box and finds the correct picture, following the instructions.
2. He puts a pencil through the hole next to the picture he has chosen.
3. He turns the card over. If the chosen hole is color coded, the answer is correct.
4. He proceeds to find the correct answer in each box.

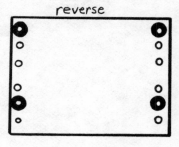

reverse

HOW MANY PAPER CLIPS?

Objective

To measure a picture with paper clips

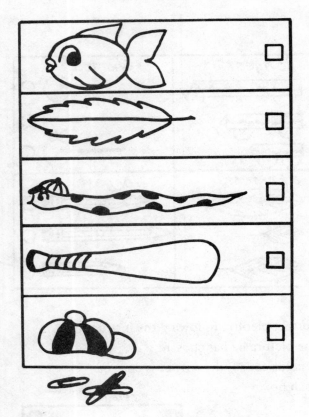

Materials

1. Folding bristol, 12" x 18"
2. Marking pen, paper clips
3. Clear adhesive plastic

Making the Game

1. Along the left side of the board, draw items (nails, tables, fish, flowers, etc.).
2. On the right side, across from each picture, draw a square.
3. Cover the board with plastic.

Playing the Game

1. The child takes paper clips and puts them together to measure each picture.
2. When he finds out how many paper clips equal the length of each picture, he writes the answer in the box opposite the picture.

MY HAND HOLDS

Objective

> To identify the number of items a child can hold in his hand

Materials

1. Duplicating master
2. Plastic margarine containers
3. Objects (small shells, macaroni, marbles, spools, paper clips, etc.)

Making the Game

1. Collect items and place grouped objects in the containers.
2. Label a duplicating master, "My Hand Holds." Leave a blank space on the upper portion.
3. Draw the items that have been collected. Make a blank line next to the picture (see illustration).

Playing the Game

1. The child traces his hand on the upper portion of the duplicated sheet.
2. The child reaches into a container and picks up as many of the grouped objects as he can.
3. He counts the handful and records the number on the blank line next to the picture of the object.
4. He continues until all blanks are filled.

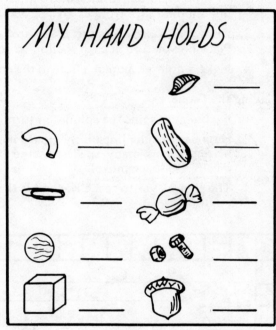

THE RULER IS KING

Objective

To measure a straight line with a ruler

Materials

1. Bristol board, 18" x 24" and 8" x 8"
2. Twelve sets of small pictures, two to a set
3. Marking pens, paper fastener, ruler, compass, paper punch
4. A playing piece for each player
5. Clear adhesive plastic

Making the Game

1. Make a track around the bristol gameboard. Each space should measure 1" x 1".
2. In random order, draw twelve lines in the center of the gameboard. The length of each line should range from 1" to 12". Label each line with a picture.
3. Draw a 7" circle on the 8" bristol square. Divide the circle into twelve equal sections. Place one picture in each section to correspond to a picture on the board.
4. Make a pointer, punch a hole in the end, and attach it to the spinner with the paper fastener.

Playing the Game

1. Each player spins the spinner in turn.
2. He measures the line on the board indicated by the picture on the spinner. He then moves his marker as many spaces on the board as the line has inches (for example, a 4" line means a 4–space move).
3. The first player to reach the end is the winner.

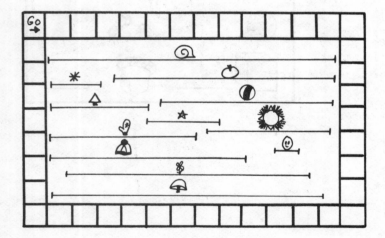

BALANCE IT OUT

Objective

To compare the weight of two sets of objects

Materials

1. Collection of small objects (macaroni, acorns, corn kernels, peanuts, nuts and bolts, marbles, etc.)
2. Folding bristol
3. Marking pens, crayons
4. Margarine containers
5. Balance scale
6. Clear adhesive plastic

Making the Game

1. Rule the bristol into 3" x 4" sections. Make task cards, following the illustration (for example: 2 marbles = __(?)__ paper clips). Cover the task cards with plastic and cut them apart.
2. Place the collections of small objects in the margarine containers.

Playing the Game

1. Place the balance scale in the center of the playing area. Place the task cards face down.
2. The player takes a task card from the top of the deck.
3. Following the instructions on the task card, he counts out the indicated object, places it on the scale, and adds the second indicated object to balance the scale. He records his discovery on the card with a crayon.

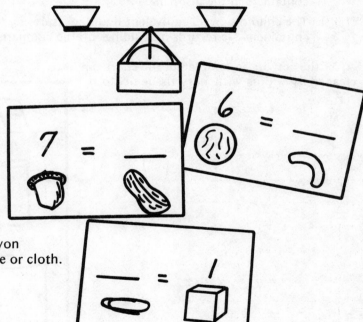

After the activity is completed, the crayon marks can be wiped off with a dry tissue or cloth.

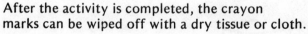

ALL WET

Objective

To understand English and metric liquid measure

Materials

1. Containers (4 oz., 8 oz., 32 oz.; 5 centileter, 1 centileter, 1 liter)
2. Pail, dishpan
3. Bristol board, 9" x 12"
4. Marking pens, crayons
5. Clear adhesive plastic

Making the Game

1. Draw a comparison chart on the bristol board (see illustration)
2. Cover the board with plastic.

Playing the Game

Set up the experiment in the dishpan to cut down on splashes.

1. Fill the pail with water (food coloring may be added, just for fun). Set the pail and the containers in the dishpan.
2. The child discovers equivalent liquid measures by filling containers with water from smaller containers. He records his findings on the comparison chart with crayon.

After the activity is completed, the chart can be cleaned by wiping with a dry tissue or cloth.

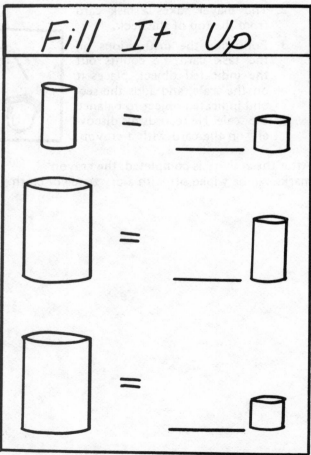

FIND THE TIME

Objective

To identify hour and half–hour on the clock

Materials

1. Bristol board, 8" x 10½"
2. Folding bristol
3. Marking pens, ruler, compass, paper punch, paper fastener, stapler
4. Clear adhesive plastic

Making the Game

1. Draw a 7" circle on the bristol board. Mark in numerals to make a clock face. Cover with plastic.
2. Make clock hands. Cover with plastic, punch a hole at the end of each hand and use the paper fastener to attach the hands to the clock.
3. On the folding bristol, rule off several 2" x 3" cards. On one side of each card, write a time (hour or half–hour). On the other side of each card illustrate the time. Cover with plastic and cut apart.
4. From folding bristol, make a 3" x 5" pocket to hold the small time cards. Staple the pocket to the back of the clock.

Playing the Game

1. The child takes the small time card from the pocket.
2. He reads the card and moves the hands on the clock to show the time. He can see if he is correct by checking the illustration on the reverse side of the time card.
3. Play continues until all cards have been used.

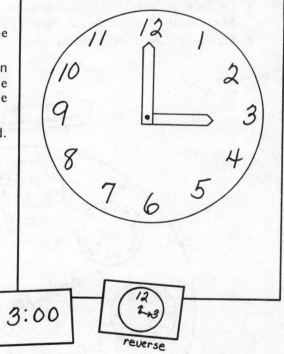

CLIP THE CLOCK

Objective

To identify hour and half-hour on the clock

Materials

1. Pizza board or circle made from bristol board
2. Marking pens, compass
3. Clock stamp (optional)
4. Spring-type clothespins
5. Clear adhesive plastic

Making the Game

1. Make circles around the edge of the circular board. Draw in the numbers of a clock and hands showing hours and half-hours, or use a clock stamp. Cover the board with plastic.
2. Using the same number of clothespins as you have made circles, write the corresponding time on each clothespin as on the clock board.

Playing the Game

1. The child spreads out all the clothespins on the playing area.
2. He picks up a clothespin and names the time written on it.
3. He finds a clock that corresponds to the time on the clothespin. He clips the clothespin to the board next to the corresponding clock.
4. Play continues until all clothespins are used and all clocks are covered.

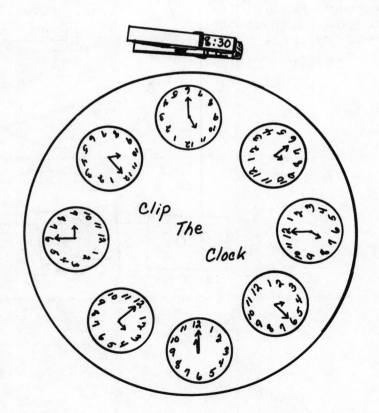

CLOCK BINGO

Objective

To recognize time on the clock

Materials

1. Bristol board
2. Folding bristol
3. Marking pens, compass, ruler, paper cutter
4. Clock stamp (optional)
5. Chips or buttons
6. Clear adhesive plastic

Making the Game

1. Rule the bristol board into 6" x 12" sections. Divide each section into two rows of four spaces each. In each section make a circle, using the compass or a clock stamp. Randomly place clock faces that show the hour. Make sure that no two rows or cards are identical. Cover with plastic. Cut apart into individual bingo cards.
2. Rule the folding bristol into 3" x 4" cards. Place the hour, either writing it out (6:00) or drawing a clock face illustrating the hour. Cover with plastic and cut into individual stimulus cards.

Playing the Game

1. Give each player a bingo card.
2. Call out the first stimulus card.
3. The players who have a match on their card will cover it with a chip.
4. The first player to cover a horizontal row wins the game.

For a higher skill, call out the hour but do not show it.

COIN CHASE

Objective

To recognize coin values

Materials

1. Bristol board, 12" x 18"
2. Coin stamps (penny, nickel, dime, quarter) or actual coins
3. Marking pens, ruler, pencil, paper fastener, paper punch, compass
4. A playing piece for each player
5. Clear adhesive plastic

Making the Game

1. Draw a track of 1¼" squares around the bristol board. Randomly mark in coin values (1¢, 5¢, 10¢, 25¢). Write GO and draw an arrow in the top left corner of the gameboard.
2. In the center of the board, draw a 4" diameter circle. Divide it into five sections. Place coin stamps (or actual coins) in each section and draw a "sad face" in the fifth section. Cover the board with plastic.
3. Make a pointer, cover it with plastic, and cut it out. Punch a hole in one end and attach it to the center of the circle with the paper fastener.

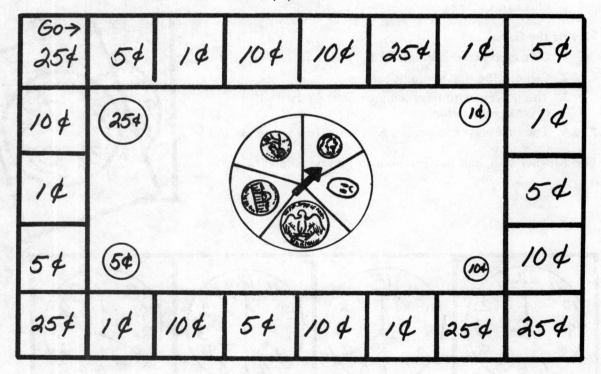

Playing the Game

1. The first player spins the spinner and moves to the nearest space that has the matching coin value.
2. Play continues in turn. If the spinner points to the "sad face," the player forfeits a turn.
3. The first player to reach the end wins the game.

MONEY MATCH

Objective

To recognize coin values

Materials

1. Bristol board, two colors
2. Coin stamps
3. Marking pen
4. Clear adhesive plastic

Making the Game

1. Mark off twenty-four 3" x 3" squares (twelve of each color).
2. On the squares of one color, place a coin stamp in the amounts to be reinforced.
3. On the other colored squares, write amounts to correspond with the stamped coin cards.
4. Cover all with clear plastic and cut apart into individual cards.

Playing the Game

1. All cards are placed face down on the playing area.
2. Each player turns over two cards, one of each color. If the coin card matches the amount on the other card, he may keep the cards and play again. If they do not match, he replaces them face down.
3. Play continues in turn until all cards are matched.
4. The player with the most cards is the winner.

HOW MUCH IS IT?

Objective

To recognize coin values

Materials

1. Eight sheets of bristol board, 9" x 12"
2. ·Store coupons
3. Marking pen, paper cutter
4. Coin stamps (penny, nickel, dime, quarter)
5. Clear adhesive plastic

Making the Game

1. Divide each bristol board sheet into six 2½" x 6" rectangles.
2. On four sheets, paste a store coupon valued from 5¢ to 40¢ on each rectangle. Cover with plastic. (Do NOT cut apart.)
3. On the remaining four sheets, place coin stamps valued from 5¢ to 40¢ to match the store coupons. Cover with plastic. Cut apart into individual coin cards.

Playing the Game

1. Give each player a coupon board.
2. Place the coin cards face down in a pile.
3. The caller turns over a coin card and names the value.
4. If a player has that amount on a coupon on his board, he takes the coin card and covers the coupon.
5. Play continues until one person wins the game by covering all his coupons with coin cards.

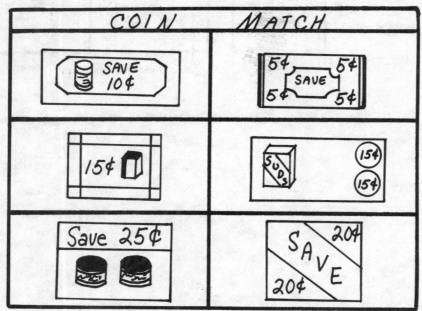

SLAP THE AMOUNT

Objective

To recognize coin values

Materials

1. Bristol board
2. Coin stamps
3. Paper cutter
4. Clear adhesive plastic

Making the Game

1. Rule the bristol into forty-two 3" x 3" sections. On each section place coin stamps. Make one quarter card, six penny cards, six cards with two pennies, six cards with three pennies, six cards with four pennies, six cards with five pennies, seven cards with a nickel stamp, and four cards with a dime stamp.

2. Cover all with plastic. Cut apart into individual cards.

Playing the Game

1. Place all the cards face up on the playing area.
2. The leader calls out a coin value (1¢, 3¢, etc.).
3. The players find the amount called and slap one hand on the card.
4. The first slapper takes the slapped card from the playing area.
5. Play continues with the leader calling another amount.
6. When all cards have been removed from the playing area, the player with the most cards in his pile is the winner.

THIRDS SORT

Objective

To assemble objects that have been divided into thirds

Materials

1. Folding bristol
2. Marking pens, paper cutter
3. Assortment of small pictures showing a single object
4. Clear adhesive plastic

Making the Game

1. Rule the bristol into 3" x 3" sections. Place a picture (or draw an object) in the center of each section. Cover with plastic. Cut apart into cards.
2. Cut each card into three pieces.

Playing the Game

1. Mix the cards and place them on the playing area (either face up or face down, depending upon the degree of difficulty you want).
2. The child will assemble the cards into whole pictures.

For a higher degree of difficulty, mix a set of "halves" with the "thirds." The player will name which picture is made of thirds and which is made of halves.

FRACTION MATCH

Objective

To recognize and match fractions

Materials

1. Four paper plates
2. Folding bristol, assorted colors
3. Marking pens, compass, ruler, protractor, paper cutter
4. Clear adhesive plastic

Making the Game

1. Following the illustration, divide one paper plate into thirds, one into halves, one into fourths, and leave one whole. Color in one section of each of the fraction plates, and color in the entire "whole" plate. Write 1/2, 1/3, 1/4, and 1 on the plates.
2. Rule the folding bristol into 3" x 3" sections. Draw a variety of shapes on the sections. Leave some whole and color the shape black. Divide the others into halves, thirds, and fourths, and color in one section on each card with black. For self-correction, write 1/2, 1/3, 1/4, and whole on the reverse side.
3. Cover with plastic and cut apart into individual cards.

Playing the Game

1. Place the plates on the playing area. Mix the cards and place them face up.
2. The player sorts the cards onto the correct plate. For a check, he can turn the cards over. They should all match values on the plate.

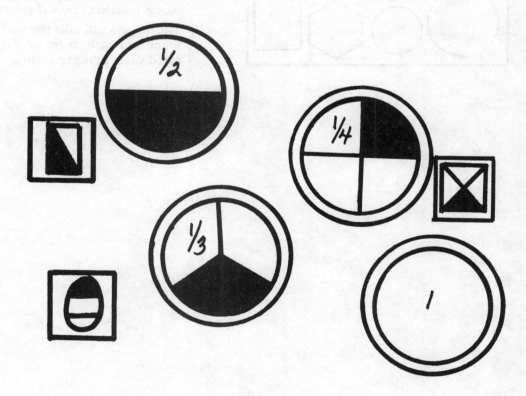

CUT–UPS

Objective

To understand the meaning of one half; that a whole is divided into two equal parts

Children should have good cutting skills before they are given this activity.

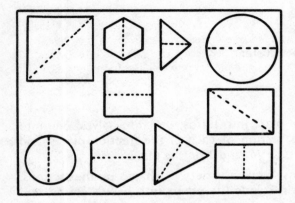

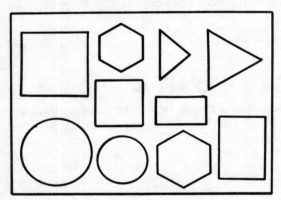

Materials

1. Two duplicating masters
2. Ruler, scissors
3. Shapes to trace

Making the Game

1. Following the illustration, draw a set of shapes on one duplicating master. Divide the shapes in half, varying the manner as often as you can.
2. On the second master, duplicate the shapes in random order.

Playing the Game

1. Pass out the first duplicated sheet.
2. Have the players lightly color in the shapes. Have them cut out each shape.
3. Discuss what a half means. Then have the players cut each shape in half and place the halves together to see if they are equal.
4. The players will take the two halves of each shape and paste them on the second duplicated sheet to make a whole.